THE CORNERSTONE OF DEVELOPMENT

Integrating Environmental, Social, and Economic Policies

Edited by
Jamie Schnurr
Susan Holtz

THE CORNERSTONE OF DEVELOPMENT

Integrating Environmental, Social, and Economic Policies

IDRC
INTERNATIONAL DEVELOPMENT
RESEARCH CENTRE
CANADA Ottawa, Canada

LEWIS PUBLISHERS
Boca Raton London New York Washington, D.C.

Library of Congress Cataloging-in-Publication Data

The cornerstone of development : Integrating
    environmental, social, and economic policies / edited
    by James Schnurr and Susan Holtz.
            p.   cm.
        Includes bibliographical references.
        ISBN 1-56670-353-0 (Lewis Publishers)
        ISBN 0-88936-842-2 (International Development Research Centre)
            1. Environmental management. 2. Environmental
    policy. 3. Sustainable development. I. Schnurr, Jamie. II. Holtz, Susan.
    GE3000.C64 1998
    363.7′05—dc21   98-34436
                                                                                         CIP

Canadian Cataloguing in Publication Data

Main entry under title:
The cornerstone of development : integrating environmental, social, and economic policies

Includes bibliographical references.

ISBN 0-88936-842-2

1. Sustainable development – Developing countries.
2. Community development – Developing countries.
3. Economic development – Developing countries.
I. Schnurr, Jamie.
II. Holtz, Susan
III. International Development Research Centre (Canada)

HC79.E5C67 1998   338.91   C98-980139-X

Co-published by
Lewis Publishers
2000 Corporate Blvd., N.W.
Boca Raton, FL 33431, U.S.A.

and by
International Development Research Centre
P.O. Box 8500
Ottawa, ON
Canada K1G 3H9

No claim to original U.S. Government works
International Standard Book Number 0-88936-842-2
International Standard Book Number 1-56670-353-0
Library of Congress Card Number 98-34436
Printed in the United States of America 1 2 3 4 5 6 7 8 9 0
Printed on acid-free paper

Contents

✧ ✦ ✧

PART IV: CONCLUDING PERSPECTIVES

Foreword

❖ ◆ ❖

The release of the Brundtland report in the late 1980s and the preparations leading up to the Earth Summit in Rio in 1992 gave prominence to the words "sustainable development." That phrase is now used to describe everything from corporate environmental performance, to the policies of the World Bank, to the priorities of Canadian government departments.

Most policies with the label "sustainable development" are in reality environmental-protection or environmental-management policies. Although environmental protection is a necessary condition, it is not sufficient. The Brundtland report's definition of sustainable development — "the ability to make development sustainable to ensure that it meets the needs of the present without compromising the ability of future generations to meet their own needs" (WCED 1988) — implies a strong social and ethical dimension. It implies the need to integrate environmental and social concerns into all major economic decisions.

In the euphoria that followed the Earth Summit in Rio, many people were excited about the opportunities for painless win–win solutions that sustainable-development strategies would offer. Although such opportunities do exist, we are now beginning to realize they're not as common as we had hoped they would be. The familiar need to make hard choices still predominates, especially in developing countries, where financial and trained human resources are in short supply.

Multistakeholder decision-making is one of the ways of making these kinds of choices. And here the Canadian experience could well be helpful. Canada's official response to the Brundtland report was to set up the National Task Force on Environment and Economy; its members were provincial environment ministers, the federal environment minister, and the chief executive officers of some of the country's largest corporations. One of its strongest recommendations was that round tables be created to integrate environment into economic policy. Subsequently, the federal government and all of the provinces established these round tables. Although many have since ceased to exist, the national round table and three provincial ones are still functioning.

Building on this round-table experience, a British Columbia commission and an Ontario commission used multistakeholder decision-making to resolve some politically difficult land-use disputes. These forums, described in Chapters

4 and 5, added a social-sustainability dimension that, for the most part, had been largely absent from the round tables.

In 1992, the Canadian government gave the International Development Research Centre (IDRC) a mandate to act as an "Agenda 21 organization," following up on the recommendations of the Earth Summit's core document, Agenda 21 (UNCED 1992). IDRC organized two early workshops to find out whether Canada's experience with multistakeholder groups could be of benefit to developing countries. Following these workshops, IDRC developed a research theme on the integration of environmental, social, and economic policies.

The essays in this book are drawn from work supported under that theme. Valuable case histories from Africa, Latin America, Asia, and Canada illustrate the pitfalls and rewards of integrating environmental, social, and economic policies. They also show how difficult it is to move from theory into practice, particularly if the tools you need for this process are absent or not well developed.

Open and clear communication must be established and maintained if diverse stakeholders and sectors are to achieve the degree of consensus that policy integration requires. And it goes without saying that the multistakeholder approach must be culturally and politically appropriate. What works in Canada or the Andean region will not work in Africa or the Philippines without major adaptations.

IDRC has completed its work with this particular program. The process and the results have had a strong influence on the way IDRC now carries out its activities. Its six current thematic areas — food security, biodiversity conservation, equity in natural-resource management, strategies and policies for healthy societies, sustainable employment, and information and communication — all promote the integration of environmental, social, and economic concerns.

Three-year program initiatives are IDRC's principal instruments for exploring a thematic area. Several of these program initiatives address the question of integration very explicitly. By organizing its support for research in this way, IDRC encourages its research partners in the South and in Canada to explore and document the most effective ways of taking into account the three imperatives — the environment, the economy, and society — that must all be respected if sustainable development is to take place.

David Runnalls
Senior Advisor
International Development Research Centre *and*
International Institute for Sustainable Development

Caroline Pestieau
Vice President, Programs Branch
International Development Research Centre

Introduction

J. Schnurr

❖ ✦ ❖

Rio+5 revealed that member states have been less than diligent in implementing Agenda 21 (UNCED 1992), the action plan for the 21st century developed at the United Nations Conference on Environment and Development in 1992. The lack of progress has placed a sobering cloud over the political and public will to respond to global environmental challenges. Although the enthusiasm and public policy support for sustainable development have been less than desired, progress has been made in understanding and documenting environmental problems at local and global levels. A better understanding of the issues, together with public pressure, has led many governments, multinational corporations, and other organizations to initiate, design, and implement sustainable-development policies and plans.

In 1992, the International Development Research Centre (IDRC)[1] was designated an Agenda 21 Organization by the Canadian government, as part of Canada's commitment to implementing Agenda 21. In response to Chapter 8 of Agenda 21, "Integrating environment and development in decision making," IDRC formulated the research theme "Integrating environmental, social, and economic policies." The theme's working-group members included experts from across IDRC's four program divisions.

The integration of environmental, social, and economic policies is clearly one of the cornerstones of sustainable development. The working group realized that a significant amount of research had already been done on the conceptual dimensions of sustainable development, as well as on the environmental impacts of economic development. However, the literature lacked documented examples and analyses of lessons learned in terms of (1) case histories of integrated approaches to policy development; and (2) the processes, procedures, and institutional arrangements needed to formulate and implement integrated policies.

---

[1] IDRC is a Canadian Crown corporation, formed in 1970 by the Parliament of Canada. Its mission, "empowerment through knowledge," is carried out by supporting research partners in Southern countries to develop practical solutions to their problems. Projects are designed to use and strengthen resident human resources and institutions and to use research networks and Canadian partnerships to foster global knowledge sharing.

IDRC's approach to policy integration

IDRC took a dual approach to its research on policy integration. One component of its approach was exploration, and the other was learning by doing. The Centre reviewed all IDRC-funded projects to identify common threads. Then it used specific case studies to explore the conceptual and practical dimensions of integrated-policy development. The aim of the case studies was to identify research areas that would in turn better serve the development and promotion of practical solutions. Several studies were commissioned in Africa, Asia, Canada, and Latin America, and four workshops were held in each region to learn how researchers and policymakers address the issue of policy integration. The outcomes of the studies and workshops formed the basis for this volume.

The concept of integrated policy

The Brundtland Commission defined *sustainable development* as development that "meets the needs of the present without compromising the ability of future generations to meet their own needs" (WCED 1988, p. 43). This definition has been interpreted in different ways, with emphasis on either the economic, social, or ecological system as the primary system and the one on which the others depend.

In Chapter 1, Robinson and Tinker set out a framework for understanding the integration of environmental, social, and economic policies. They suggest that defining the ecological system as the primary system can be seen as imposing a particular, elitist view about environmental issues and blocking progress and human development, particularly in the South. On the other hand, defining the socioeconomic system as the primary system can be seen as providing "a veneer of environmental respectability for what is really continuing, non-sustainable economic growth" (Robinson and Tinker, this volume, p. 21). Robinson and Tinker argue that "to escape from this deadlock, we need to forge imaginative new approaches that recognize and integrate ecological, social, and economic conditions and goals" (Robinson and Tinker, this volume, p. 21).

The assumption of the studies in this volume is that no one of these systems is more important than the others. Rather, the emphasis is on movement toward sustainability in all three systems. The process is seen as one of ongoing reconciliation between systems' imperatives, and this involves identifying, assessing, and effectively managing the trade-offs between the imperatives in any given context. The aim of the process is steady progress toward human–ecosystem well-being.

The country case studies illustrate a cross section of unique experiences in the private and public sectors, including work done by nongovernmental organizations (NGOs). This experience includes the application of various tools, mechanisms, and institutional arrangements to integrate policy development. The findings also identify challenging issues. Is integration enough to ensure sustainable development? In what context is integration appropriate? From a governance perspective, how and by whom should integration be managed?

Not surprisingly, the emphasis on policy integration varies by region and its respective cultural and socioeconomic conditions. In Africa, the emphasis tends to be on social development, the relationship between economic and social policies, and the trade-offs between poverty and environmental degradation. The paper by Intal (Chapter 9) points out that the recent economic growth and prosperity of East Asia have pressured governments and civil society to address the environmental implications of growth. Owen's paper (Chapter 4) and Penfold's paper (Chapter 5) reveal that, in Canada, some interesting lessons have been learned about using multisectoral decision-making processes and about identifying appropriate roles for the various players in integrated-policy development and implementation.

Barriers to integration

Those who are working to integrate policies invariably face significant barriers. The cause and effect relationships among environmental, social, and economic systems are complex, interactive, and hard to unravel. Government agencies, corporate departments, and research and academic institutes are typically set up according to discrete sectors and disciplines, each with its own interests (and interest groups), virtually ensuring policy segregation. Our political economy discounts the future value of human development, natural resources, and ecological processes in exchange for short-term economic development. We find ourselves short of experience with the analytical tools and decision-making processes needed to identify, evaluate, and manage the inherent trade-offs between objectives.

In general, integration is seen as a good thing. However, on the conceptual level, the term has been overused and its appropriateness and practical implications are not well enough understood. The implications of using integration as a tool of analysis are different from those of using it as a tool for implementing policy. First, in analysis, introducing the concept of integration bears the risk of reducing analytical rigour and increasing the chances of superficiality. Second, in attempts at cross-sectoral analysis, one is forced to higher levels of abstraction, and this may divorce the analysis from the challenges of the "real world." Third, in integrating policies analytically and deriving policies based on integration, we are stepping beyond the role of analysts and more into political roles, proposing a concept of integration rather than presenting alternatives and suggesting how they work under various scenarios.[2] It is important that these limitations be carefully considered when one is proposing integration as an approach.

---

[2] Brooks, D. 1995. IDRC and INTESEP: a sceptical view. International Development Research Centre, Ottawa, ON, Canada. Unpublished manuscript.

What sort of integrated policy?

Whether integration is used as a tool for analysis or as a way of guiding a process, one way of confronting the problem of complexity is to define the different levels at which integration should take place — local, regional, national, or international. Another strategy is to approach policy integration from an ecosystem perspective, such as from the point of view of fluvial or watershed regions or bioregions based on vegetation.

There are also varying degrees of integration. A sectoral policy that is sensitive to other sectoral policies or issues could be considered to have one degree of integration. Legislation that requires social- or environmental-impact assessments (EIAs) of development projects or end-of-the-pipe abatement technologies for industrial production systems are more advanced forms of integration. Even deeper degrees of integration involve market-based instruments, green or socially responsible procurement measures, and various types of partially voluntary arrangements to make environmentally and socially responsible management a priority for government, industry, and citizens. Strategic environmental planning, life-cycle assessment, and integrated impact-assessment techniques are other tools that can foster forms of deeper integration.

Degrees of integration occur incrementally along a continuum, as shown in DePape's case study of a Canadian electric utility (Chapter 3). A series of legislated EIAs created awareness of the environmental impacts of building hydro lines on a preselected site. The learning that took place during the assessments and the desire to apply the new knowledge eventually led to changes in the utility, which instituted a strategic environmental-planning process and self-directed assessments. In the end, the utility introduced new management practices for assessing alternative sites for its transmission lines.

Coordination, participation, and governance systems

Policy integration requires coordination and collaboration in designing, planning, and implementing, to establish clear objectives and divisions of responsibility. More advanced degrees of integration require more sophisticated forms of communication, decision-making, and organizational behaviour. Mechanisms and tools such as multistakeholder forums and user-friendly information systems give people an opportunity to provide input into policy decisions. As well as contributing to informed decision-making, the process helps policymakers understand the socioeconomic and ecological contexts in which they work and also helps all stakeholders appreciate the trade-offs entailed for any given policy decision.

The Indian case study by Vyas (Chapter 8) and the Peruvian case study by Soberion (Chapter 10) show that, whether one is using multistakeholder forums, a *mesa de concertación* (round table for concerted action), or some other mechanism, the success of any participation strategy depends on the prevailing political, social, and cultural conditions. Systems of governance that anticipate societal responses to various integrative measures and accommodate the policy objectives of a range of stakeholders and sectors are crucial. In this perspective, governance means the inter- and intraorganizational arrangements, decision-making processes, incentives, and disincentives through which governmental and nongovernmental actors — including the public, communities, and the private sector — influence decisions about societal priorities and resource allocations. Governance goes beyond formal institutions of government and includes the significant role of nongovernmental actors in policy formulation and implementation, particularly in developing countries.

Conflict and learning

Despite the best coordination and participation strategies to balance conflicting objectives, often some degree of conflict is unavoidable. Success in managing conflict lies in structuring the decision-making process so that it involves affected parties' representatives in the design and evolution of the process, as well as in the negotiations on substantive issues. Interest-based negotiation, as described in the Owen case study (Chapter 4), is one example of a structured, deliberate attempt to cooperatively achieve an outcome by accommodating rather than compromising the interests of everybody concerned.

In structured multistakeholder negotiation, learning is fostered by adopting decision-making guidelines, communication rules, and process steps. Learning can also occur without specific structures. In the case of the Canadian hydro utility (Chapter 3), legislation, along with encouragement from management and an interdepartmental committee, prompted line departments to learn from their experiences and develop more effective integrative tools.

Perhaps the most unique aspect of this collection of studies is the way it links learning theory and practice to the practice of conflict resolution and policy integration. Chapter 2, "Learning and policy integration," by Bernard and Armstrong, identifies several learning principles and describes their role in promoting and fostering integration. For example, learning can best be encouraged by allowing interested parties to jointly define rules for communication and negotiation and have equal access to information; by creating incentives for risk taking; and by allowing a margin for error. Other positive elements in a learning process are delegation of responsibility; and a willingness and the ability to capture and build on unexpected results. Learning can take place without specific structures if strong incentives and disincentives are in place.

The role of knowledge and research

Central to the resolution of conflict and learning is the availability of, and access to, information and knowledge. The Ghanaian case study by Anderson (Chapter 6) and the Kenyan case study by Syagga (Chapter 7) highlight the critical importance of relevant and accurate information and knowledge in informing decision-making and stimulating learning. Often this knowledge is generated by NGOs, as was the case in the Ghana experience. The availability of information depends largely on research, which has a valuable role to play, given the complex challenge of integrating environmental, social, and economic conditions and goals. Collaborative approaches, such as inter- and multidisciplinary research, appropriately applied, offer useful methodologies for addressing problems from an integrated perspective.

The most pressing role for research is not to further explore the concept of integration but to determine the appropriate role of integration in reconciling the environmental, social, and economic concerns in a given context; and, from a governance perspective, to determine how and by whom integration should be managed. Researchers can help identify the range of trade-offs that are necessary in any given scenario and identify and assess the options available to decision-makers. Researchers can help develop and test tools and methods for identifying and assessing policy options. Researchers can also help develop and assess the role of various planning and management tools and institutional arrangements needed for policy formulation and implementation.

One key finding of this research is that integration hinges on the ways trade-offs are evaluated and managed. Political institutions and policy-making processes also need to have the flexibility to promote and foster integration when appropriate. Ultimately, policy integration unleashes processes with unpredictable outcomes. A variety of stakeholders may be relevant in any given context, affecting both the substance and the process of policy integration. If the process is successful, inputs may arrive from diverse sources, suggesting several possible ways to meet the goals of equity and sustainability; if not, the scenarios are likely to fail to meet even minimal standards. The aim of this volume is to illustrate the conceptual and practical challenges in integrating environmental, social, and economic policies to ensure sustainable development.

PART I

THE CONCEPT OF
POLICY INTEGRATON

Chapter 1

Reconciling Ecological, Economic, and Social Imperatives

J. Robinson and J. Tinker

◇ ✦ ◇

Three challenges

As we move toward the 21st century, human institutions, from the local level to the global level, are facing a range of ecological, economic, and social challenges.

It is becoming clear that much of our industry and agriculture and our use of renewable and nonrenewable natural resources are unsustainable. Two examples suggest the global scope of the problem:

✦ Humans may now directly and indirectly appropriate about 40% of the total photosynthetic product of the planet (Vitousek et al. 1986; for marine resources, see Pauly and Christensen 1995). This will likely stringently limit future growth in human consumption.

✦ If the current global population consumed resources at the same rate as the average Canadian, two additional planets would be required (Rees and Wackernagel 1994; see also Turner et al. 1990; WRI 1994; WI 1995).

Such calculations suggest that global carrying capacity will soon be exceeded, if it hasn't been already, and that global adoption of industrialized countries' rates of consumption and production would simply be untenable.

Economically, change is now extremely rapid, including the disappearance of centrally planned economies; the powerful trend toward the use of market forces and market-based policies throughout the world; global economic integration, driven by trade liberalization; and the emergence of a global capital market, characterized by financial flows that dwarf flows of traded goods and services.

9

These developments have in turn had a number of effects:

+ Increased economic interdependence among nation states and reductions in national economic sovereignty;

+ The emergence of global corporations and financial institutions whose activities cannot now be effectively regulated by governments;

+ Highly mobile international trade and investment flows, which are felt to limit nations' freedom to raise taxes for social programs;

+ Increasing pressures to maintain international competitiveness;

+ Pressures to reduce the size of the public sector, to reduce (or at least not increase) taxation (especially direct taxes), and to reduce deficit financing and public debt;

+ Growing structural unemployment in many industrialized countries;

+ A rising and unacceptable number of people living in absolute poverty; and

+ Large income disparities between richer and poorer countries and between rich and poor within both industrialized and developing countries.

The causes of these problems are the subject of much debate, as are the most promising remedies; in some cases, the debate is about whether these phenomena are problems at all. But current economic conditions are clearly unsustainable for a significant proportion of the world's population, in developed as well as developing countries.

Governance and other social structures are also under unprecedented stress. In many market-oriented industrial societies, the system of governance is viewed with growing distrust, a sense of alienation, and even distaste. This is coupled with the failure of governments to address basic social issues, such as crime, drugs, poverty, unemployment, and homelessness, in ways that command public support. Such alienation may grow as public demands to cut taxes and reduce debt conflict with the desire to maintain social and environmental programs. The overall effect is a decline in civil society and, in many inner-city neighbourhoods, a descent toward lawlessness and ungovernability.

In the former centrally planned economies, fragile structures of governance are often barely surviving the stresses and social problems accompanying the transformation to a market economy. In the developing world, poverty, rapid population growth and displacement, the replacement of a subsistence economy, other forms of economic development, and massive environmental impacts are being managed with only mixed success, perhaps best in parts of Asia and worst in parts of Africa. The major challenge in many former command economies and many developing countries faced with a rapid decline or even collapse of traditional value systems is to enlarge and strengthen a stable civil society, which at present is only embryonic. Without a stable society, the trust and public self-confidence needed for participatory governance are limited.

The end of the Cold War and the winding down of the superpower nuclear arms race have not ended high worldwide levels of military expenditure but have revealed new instabilities. Tensions between different ethnic groups and the demand for subdivision of existing states often lead to armed conflicts that international mechanisms fail to resolve. At the same time, social cohesion is declining in many if not most societies. Societal and cultural dislocation, fueled by the globalization of communication, is endangering the existence of many small cultures (especially those of indigenous peoples) and may threaten the health or integrity of many more; even cultures that feel invulnerable are experiencing a decline in sense of community.

The extent of these problems illustrates a form of social unsustainability: in many parts of the world, we may have exceeded the "carrying capacity" of our current cultures and governance systems (for more details, see UNDP 1994; UNICEF 1994; World Bank 1994).[1]

Interconnections

Since the 1960s, and even more so since the mid-1980s, the inadequacy of policies that fail to recognize the interconnectedness of ecological, economic, and social challenges has become apparent. In fact, the interaction of these challenges often reinforces their negative impacts. For example,

+ A legacy of environmental mismanagement is now seen as one of the most severe economic burdens on the countries of the former Soviet Union;

+ Internationally agreed targets for reducing greenhouse gas emissions are not being reached because governments are unable or unwilling to accept the economic cost;

+ Addressing the mismanagement and environmental deterioration of agricultural land often proves difficult or impossible with current land-tenure and other socioeconomic structures;

+ Growing trade liberalization and structural readjustment are in many countries seen as (or blamed for) being directly responsible for reductions in health, education, and other social programs;

---

[1] Rapid population growth will continue to have major impacts on society, the market system, and the biosphere, although completed family size is now falling in some developing countries. Rapid population growth may well end by the close of the 21st century. However, the level at which global population stabilizes (and its geographical distribution) will have a massive influence on the feasibility of the dematerialization and resocialization strategies outlined in this chapter. This topic deserves an explicit treatment, which we do not provide here.

+ Increasing human demand for declining or damaged natural resources, such as water, agricultural land, and pasture, is a major cause of social and ethnic tensions, often leading to armed conflict; and

+ Economic inequality within societies, especially when it appears to be connected to increasing globalization of the world's economies, is reducing social cohesion and making it more difficult for people to accept macroeconomic change without social tension.

Perhaps most fundamental is the fact that in the last half century, economic objectives have been having a major influence on the other two systems, whereas the ability of the social system (via the state) to influence the economic system has been declining. On the other hand, there is also evidence that other social influences on the market are growing (via consumer and environmental groups, for example) (Drucker 1995; Elkins 1995). On the whole, however, in many parts of the world, global economic integration clearly seems to be connected with social and cultural fragmentation or even disintegration. For example, increasing "secession of the elites" from a society of declining mutual obligations and growing economic inequities has been linked to continued economic globalization (Schrecker 1998). This is not merely a matter of the direct economic and employment effects of, say, trade liberalization or structural-adjustment policies. Arguably, the perceived as well as the actual loss of national economic control associated with global economic integration is a major factor in increasing social tensions and in reinforcing a desire to build a sense of community through local sovereignty and separatist groups of various kinds. For example, Ignatieff (1993) has argued that the recent increase in ethnic nationalism around the globe is closely tied to a decline in civic nationalism (the collective sense of security and trust in national authority and institutions), although Elkins (1995) has expressed a more positive view of the unbundling of the nation state. Perhaps even more disturbing, though, are the causal connections that Homer-Dixon (1991) found among environmental degradation, economic development, population growth, refugee movements, and war. Problems of national and subnational security, the arms trade, and increased militarization are connected in complex ways with economic, social, and ecological factors.

Ecological, economic, and social unsustainabilities are mutually reinforcing in a least two ways. First, they have direct effects on each other. Second, addressing any of these issues in isolation, without considering their interactive effects, can give rise to unanticipated higher order consequences in the other realms, which cause problems of their own or undercut the initial policies. For example, raising energy prices significantly to reduce energy emissions will disproportionately affect poorer citizens, thus increasing income disparities and contributing to social unsustainability.

Poorer citizens spend a greater proportion of their income on energy in countries where most energy requirements are met through purchase of commercial fuels. In some developing countries, however, commercial fuels are used mainly by wealthier urbanites, whereas poorer rural citizens use nonmarket sources of energy. In such cases, increases in the price of commercial energy sources will effect mainly the richer citizens, with quite different equity (not to mention economic and sociopolitical) consequences.

The extensive interactions among the ecological, economic, and social systems suggest that none of these three groups of issues can usefully be addressed in isolation. This chapter proposes a single conceptual framework for considering ecological, economic, and social objectives. It suggests how these different objectives might be reconciled, as well as the need for a new analytical framework to evaluate the ability of local, national, and international policies to reconcile these three types of goals.

The arguments of this chapter are preliminary. Much work remains to be done in a number of fields for these arguments to be fully fleshed out. They represent an initial attempt to provide an integrated look at these issues, with the hope that any response to the positions taken here will help us develop the arguments further.

The present conceptual framework: trifocal vision

One of the main obstacles to developing a common conceptual framework for ecological, economic, and social problems is the little consensus on how the three systems relate. The three relevant disciplines provide us with very different views of the world that are difficult if not impossible to reconcile in a single mental image. Constructing an integrated understanding from these three is like trying to see with trifocal vision; making effective management decisions is even harder.

— Environmentalists often insist that both the economy and human society are subsets of the global ecosystem and must obviously be subject to constraints based on ecological limits. From this point of view, continued economic and population growth will cause breakdowns in ecological life-support systems and exceed the carrying capacity of the biosphere, which in some respects has already happened. Environmentalists stress that nonrenewable resources are finite; that renewable resources can be (and often are) overexploited, leading to collapse; and that losses of biodiversity are irreversible.

— Economists tend to argue that both ecological and social factors can be expressed in economic or market terms (for example, by cost–benefit analysis) and that ecosystems and society, by definition, lie within the global economy. Many believe that market-driven substitution is an answer to depletion or exhaustion of renewable or nonrenewable resources and deny that there are any real or immediate global limits to economic growth. Some argue that market solutions are available for all ecological and social problems.

Social and political scientists also tend to see their discipline as fundamental. They tend to address environmental and economic issues as aspects of
human social issues. Although sociologists are often sympathetic to environmental concerns, they argue that neither ecology nor economics can adequately
address issues such as social justice and equity (including intergenerational
equity).[2]

All these insights are valuable. A more or less convincing case can be made
for saying that the biosphere encompasses the market and human society; that
economics can address environmental and social concerns; or that the economy
and the environment are best regarded as cultural subsets, or constructs of
human society. However, ecologists, economists, and social scientists are all generally resistant to the more extreme claims of those in other disciplines, that the
ecosystem, the economic system, or the social system has fundamental primacy
over the others. Such a view is not widely accepted outside its related discipline,
nor does any such view seem likely to become so.

We argue that it is more fruitful to consider the biosphere, the market,
and human society as three interacting "prime systems" that share many common characteristics but are equivalent in primacy and importance. Some of the
implications of this approach are explored in the remainder of this chapter.

Toward a new conceptual framework:
the three prime systems

The three prime systems, interconnected, overlapping, and coequal, as defined
above, are

- ✦ The biosphere, or ecological system (the planetary biogeochemical system *sensu stricto*; it is not a closed system, as it depends on energy
  exchange within the solar system);

- ✦ The economy, the market, or economic system; and

- ✦ Human society, the human social system, which includes the political
  system (governance), the social system (family, communities, etc.), and
  cultures.

Of course, this division into three prime systems is to some extent arbitrary. Many other divisions are possible. Some people, for example, might prefer to make human population a fourth system (a subset of our social system);
others might consider the technosphere, or industry, a fifth (a subset of our

---

[2] Interconnections of social (or sociopolitical) and economic factors have long been studied
and accepted, as suggested by the term *socioeconomics*. Recognition of the interrelationships of ecological factors and economic and social factors is more recent, giving rise to newer concepts, such as
those of *human ecology* and *industrial ecology*.

economic system). Astronomers would regard all these systems as subsets of the solar system or the universe.

The three prime systems postulated here can be visualized as an equilateral triangle. An alternative model would be an equilateral pyramid (a tetrahedron), with the three grounded corners symbolizing the biosphere, the economy, and society and with the apex symbolizing politics. This alternative conceptual framework emphasizes that politics is the system or tool by which we attempt to manage all three other systems. We prefer to treat the sociopolitical system (including its cultural component) as a single system, largely because no clear distinction can be made between the political system — governance — and the social system in which it is rooted.

We have adopted the equilateral framework because it seems an intuitively reasonable structure, reflects common usage, and leads to useful insights into human problems, responses, strategies, and policies. We do not claim that ours is the correct framework and that other frameworks are wrong, any more than a dictionary is logically preferable to a thesaurus. A conceptual framework is no more than a mental construct; its validity derives primarily from its utility.

As the following subsections show, the prime systems display, to a greater or lesser extent, a number of common attributes: having subsystems and being affected by globalization; being self-organizing; being capable to some extent of responding to stress; and having more or less inflexible outer limits beyond which the system will collapse.

Subsystems and globalization

The prime systems can each be divided into subsystems of varying size, both spatially and sectorally. Also, all three systems seem to be undergoing changes related to globalization.

Although the planetary biosphere is one ecosystem, it can be subdivided spatially: each wood, marsh, or farm can be considered a distinct ecological community. Larger units may have a geographical basis: Canada, for example, has been divided into 15 ecozones. The global ecosystem may also be subdivided sectorally into broad and often noncontinuous biomes, such as subarctic tundra, tropical moist forest, and coral reef. The biosphere has, of course, always been a global system, but anthropogenic impacts on it (those derived from human activities) are unquestionably undergoing a rapid globalization: water pollutants, for example, were once confined to rivers downstream of settlements, but they are now found throughout the oceans.

The economy, or market, can also be divided spatially or sectorally. We may consider the economy of a country, of one of its regions, of a town or village, or of a small community or family. Alternatively, we may consider the market in terms of a particular sector or commodity, such as copper, timber, sorghum, or recreation.

The market is also undergoing rapid globalization, driven both by communications technology (causing the progressive global integration of the banking system and of the stock, currency, and commodity markets) and by political pressures in favour of free trade. The market was once regulated locally (for example, by village markets or town guilds) or nationally, but increasingly there is also some form of regulation at the regional level (for instance, the European Union, the North American Free Trade Area, the Association of Southeast Asian Nations [ASEAN]) and globally (the General Agreement on Tariffs and Trade, the new World Trade Organization, and the International Monetary Fund). Moreover, globalization of the world's economies — that is, increased trade, investment, and communications — tends to reduce the relative importance of national economic decision-making. The effect of these phenomena is a weakening of national economic sovereignty, except perhaps for the very largest states.

Society, too, may be subdivided, with the basic unit remaining that of the family (currently shifting in many places from the extended to the nuclear). Political subdivisions are largely spatial, from the city or province, via the state, to the region (for example, the European Union and ASEAN) and the globe (for example, the United Nations and its agencies). Sociocultural subdivisions are more often sectoral (religious, linguistic, cultural, recreational, special interest, and the like). Increasingly, social links tend to transcend geographical boundaries.

Again, communications technology is leading to a globalization of society (the global village) and to a far more rapid rate of change in societal structures than in the past: the current abandonment of Soviet-style communism and the rise of Islamic fundamentalism, for example, are occurring substantially faster than did the decline of feudalism or the Reformation in Europe. In physical terms, transport technology allows more rapid and substantial transfers of population than in the past, with the new phenomena of economic and environmental refugees added to the more familiar ones of economic migrants and political refugees. Paradoxically, this social globalization seems to be connected to various forms of social and cultural fragmentation and localization.

Self-organization

A second characteristic common to all three prime systems is that they are to a considerable extent self-organizing. The biosphere, the economy, and human society follow certain patterns, which ecologists, economists, and social and political scientists study, explain, and even try to predict. But the ability of humans to influence or adjust the three prime systems in a desired direction is limited. Observation of each prime system suggests that it operates at least partially according to internally derived and inherent principles in ways that we cannot fully understand or control.

Response to stress: system change

A third set of characteristics common to all three prime systems relates to their capacity for change: diversity, stability, resilience, and self-organization all change as a result of stress. The biosphere, unlike the other two prime systems, can exist independently of human activity, but it is increasingly intertwined with the other two in practice, and all three systems exhibit some similar behaviours.

The biosphere, the economy, and society all have a certain stability, but all three incorporate considerable diversity, are in a dynamic rather than static equilibrium, and are subject to continual change. In each case, diversity appears to be related to both stability (limited change over the long term) and resilience (the ability to absorb and adapt to stress). However, the belief that a simple, homogeneous ecosystem is by definition less stable or less resilient than a complex, heterogeneous one is ill supported by observed data (Holling 1986; Kay and Schneider 1994).

Each prime system is affected by stress created within or outside the system. The sources of these stresses include new technology and ideas; anthropogenic stress generated on a global scale is increasing in all three prime systems.

The stress-response time of each prime system varies with the stress and with the system. Oil prices, for example, can change within minutes if there is news of a threat to supply, and both society and ecosystems can also react to certain stresses very quickly. Other responses can be much slower, lasting years or decades; for example, the impacts on metropolitan Spain and Portugal of the collapse of their colonial empires were spread over more than a century, and the response of global temperature to greenhouse gas emissions is thought to be measured in many decades.

Each prime system shows considerable resilience, that is, the capacity to absorb stress without noticeable change (a capacity for buffering). But each prime system changes when stress cannot be absorbed, or when its resilience or buffering capacity is exceeded. Moreover, change in response to stress is a crucial component of long-term system stability; for example, wildfires started by lightning are now seen as an integral part of the stability of prairie and boreal-forest ecosystems. Such change is often incremental and may well be largely reversible once the stress is removed. More severe stresses can cause nonincremental change, which seems usually to be evolutionary and unidirectional (that is, irreversible) and may increase or decrease diversity, homogeneity, stability, and resilience.

In all three prime systems, major stress can cause a discontinuous change of state, a nonlinear flip into a significantly different state. This is like the response of a gas to increased pressure: first, a steady reduction in volume and then a sudden liquefaction. In the economy, society, or the environment, such a change of state is often, perhaps usually, irreversible. Pollution in the North American Great Lakes, for example, caused a major shift in fish species that was not reversed when pollution levels were reduced (Regier 1995). Clearing

Amazonian rainforests on laterite soils and replacing them with low-grade pasture is a change that cannot be reversed (at least, not for millennia) by attempting to recreate the original forests. Similar phase changes in society, such as the overthrow of the ruling class in the French and Russian revolutions, seem irreversible. Discontinuities in economies, such as the 1929 Wall Street crash and the industrialization process, also seem to be irreversible. In general terms, major changes of state occurring in a prime system in response to stress are not followed by a return to the *status quo ante* after the stress has been removed.

The interconnections among the three prime systems imply that changes in response to stress are rarely confined to one system. For example, where a first-growth forest is replaced by a species-poor, second-growth forest (an ecological response) after clear-cutting or burning, the consequences will also be felt by local forestry-based human communities and industries (economic and social responses). Not all of these responses are local: ecosystems are connected by global biogeochemical systems; economic systems, by increasingly global markets; and social systems, by a growing telecommunications web.

Accurately predicting system change in response to stress, especially changes in one prime system caused by stresses on another, requires greater knowledge than we currently have. Such change often occurs in counterintuitive ways. For example, in the early stages of the AIDS pandemic, it was feared that in countries severely affected by a disease that primarily affects people in their sexually active and therefore economically productive years, changes in the mortality and morbidity rates would significantly increase the dependency ratio, or the number of nonproductive persons that each working person has to support. The economic and social consequences of such an increase would clearly be harmful. In actuality, this intuitive expectation appears unrealized in places like Africa, where transmission is primarily heterosexual. Here, vertical transmission (perinatally, from mother to child) occurs as well as horizontal transmission (between sexual partners, or by contaminated blood). Therefore, it seems probable that the rise in infant and child mortality from AIDS counterbalances deaths among productive age groups, holding the dependency ratio roughly stable (Anderson 1993).

The desirability of change

It can be argued that another common characteristic of all three prime systems is having more or less inflexible outer limits, beyond which the system will collapse. For example, it is often claimed that the biosphere's limits may be exceeded by the continued emissions of greenhouse gases, leading to the total breakdown of the system. Similar system collapse has been predicted for the economy if major Third World debtor nations discontinue repayments to Western banks. It can be argued that a equivalent collapse of the social system has recently occurred in Somalia and Rwanda.

Whether such analyses are correct depends on how one defines *collapse*. If system collapse is defined anthropocentrically as being a major change with unpleasant or fatal consequences for humans or even (in the case of the biosphere) for the human species, then such a fundamental breakdown seems possible. But from the point of view of the system, this is not a collapse, merely a major flip, or phase change. It is important to be clear about what is collapsing, what is adapting, and what remains unchanged. For example, the biosphere has in the geological past gone through substantially warmer periods. If it is subjected to such stresses in the future, it will most likely adapt to them too. Many species — and conceivably the human species — would probably become extinct; nevertheless, the biosphere itself would not be destroyed. The aftermath of World War II included traumatic changes to the German economy and society, which did not destroy them. Although civil society has changed radically (and painfully) in Somalia and Rwanda, as in any community involved in civil war or revolution, social structures have adapted and changed, rather than being destroyed completely. We're not saying that efforts to avoid disasters should be neglected but that consideration should also be given to the potential for renewal, adaptation, and innovative response.

As all three prime systems are largely self-organizing, they are highly unlikely to collapse. Gradually or suddenly, all such systems adapt in response to stress. The question is whether such change is desirable. A radical phase change may even be desirable and lead to a more stable system. In the long run, was the French Revolution of 1789, along with the Terror that followed it, beneficial or harmful to the French people? The conversion of much of Java from rainforest to terraced rice paddies put an agricultural system in place of a natural ecosystem, but this agricultural system has remained stable for many centuries and has supported a far more numerous human population than the rainforest ever did.

System change may be evaluated using two partly overlapping yardsticks:

+ Whether the direction, quality, and rate of change are beneficial or harmful to humanity (over various time scales or to individuals or defined human communities); and

+ Whether such change is ethical (which can, if desired, include consideration of intergenerational equity and of other species).

In the last resort, both yardsticks are variants of desirability and are value based, rather than being purely objective. This means that decisions about whether changes are beneficial are not scientific or technical but political. Scientific analysis can tell us about the potential consequences and likelihood of various changes, but decisions about their desirability and what should be done to avoid or adapt to them must be made in the political arena.[3]

---

[3] This discussion has focused on the commonalities among the three prime systems. There are also, of course, many important differences. Perhaps the most fundamental is that the economic and social systems depend on the existence of humans, whereas the biosphere does not. Given sufficient stress, the planetary ecosystem could alter so radically that it no longer supported our species.

Managing system change: sustainable development and its limitations

The most widely accepted basis for understanding the interdependence of eco-logical, economic, and social challenges has been the concept of sustainable development, which grew out of the ideas of ecodevelopment in the 1960s and 1970s (Sachs 1984) and the UN Conference on the Human Environment (held in Stockholm in 1972). During the 1980s, sustainable-development thinking was further elaborated in the *World Conservation Strategy* (IUCN 1980) and the work of Brown (1981) and others, such as Clark and Munn (1986). The term *sustainable development* reached public prominence with the report of the World Commission on Environment and Development, often referred to as the Brundtland Commission, in the late 1980s (WCED 1988).

The most widely used definition of sustainable development comes from the Brundtland Commission report (WCED 1988, p. 8): "Humanity has the ability to make development sustainable — to ensure that it meets the needs of the present without compromising the ability of future generations to meet their own needs." The Commission report also stated (p. 9) that "sustainable devel-opment is not a fixed state of harmony, but rather a process of change in which the exploitation of resources, the direction of investments, the orientation of technological development, and institutional change are made consistent with future as well as present needs."

Fundamental to the Commission's position were the views that sustain-able development is a global issue; that poverty and environmental concerns must be addressed together; that significant improvements in the material stan-dard of living of developing countries are a precondition to global sustainable development; and that considerable opportunities exist for improving environ-mental quality and human development through technological development and institutional reform. In a famous and controversial proposal, the Brundtland report called for a 5- to 10-fold increase in gross world economic output, to meet the development needs of the poor and to provide the wealth and technological advances required to address ecological problems.

The report of the Brundtland Commission marked the start of a series of activities and events at international, national, and subnational levels. A host of sustainable-development policies, boards, commissions, and round tables sprang up. In the private sector, organizations such as the Business Council for Sustainable Development and the International Chamber of Commerce, as well as many corporations, developed sustainable-development principles and poli-cies. These developments coincided with an upsurge of public interest in envi-ronmental issues in many countries (Dunlap et al. 1993) and perhaps reached their culmination in the Earth Summit (and the accompanying NGO Global Forum), held in Rio de Janeiro in 1992. At that conference, attended by more heads of state than any previous meeting, several documents were signed, including Agenda 21 (UNCED 1992), an international action plan for sustainable development; the Framework Convention on Global Climate Change; the Biodiversity Convention; and the Statement on Forest Principles. Although public interest subsequently declined and governments failed to pursue these initiatives aggressively, many countries, corporations, and other organizations continue to develop sustainable-development policies, although these rarely manage to fully integrate all three prime systems.

The principles and practice of sustainable development have also been criticized. The concerns stem from the different views of those engaged in related public-policy issues. As the Brundtland report pointed out, any attempt to achieve sustainable development must address a number of economic questions, including how and what is produced and consumed and how wealth and prosperity are generated. Many people with strong ecological values, especially those with a belief in finite ecological limits to growth, also consider such values as standing in opposition to the economic priorities that drive our societies toward ever greater levels of environmentally destructive production and consumption. From this perspective, ecological and economic goals are locked in conflict, and each can be satisfied only at the expense of the other; in other words, more economic growth leads to environmental collapse, whereas no economic growth leads to economic collapse. A contrary view, rooted in the belief that biophysical limits are either distant or subject to technological mitigation, is that both the aspirations of the well-off and the needs of those living in poverty demand continued economic growth; indeed, those who hold this view argue that such growth is needed to pay for the implementation of any strong environmental and social policies.

In this context, many advocates on both sides have been suspicious of the concept of sustainable development, which (as argued by Brundtland) seems to imply that the world could and must have both continued economic growth and ecological sustainability. From one perspective, this concept seems to provide a veneer of environmental respectability for what is really continuing, nonsustainable economic growth. From another perspective, it seems to impose a particular, elitist view about the overriding importance of certain environmental issues and thereby blocks progress and human development, particularly in the South (Guha 1989).[4]

To escape from this deadlock, we need to forge imaginative new approaches that recognize and integrate ecological, social, and economic conditions and goals. For example, merely imposing ecologically based constraints on economic behaviour is certain to be insufficient. Not only would such constraints continue to be resisted by powerful interests, but they too often represent an end-of-pipe approach that treats ecological solutions as an add-on, to be

---

[4] This typology grossly simplifies a complex set of positions. In particular, it fails to address the significant differences among positions articulated in different parts of the world, such as in developing countries or rapidly industrializing economies. Yet, we believe it captures a critical distinction that runs through much of the overall debate.

incorporated after the fact and only insofar as required. Moreover, such constraints scarcely even begin to address the social or economic problems caused by globalization.

A clearly preferable alternative would be to integrate environmental concerns at a deeper level, that at which ecologically sound reasoning is also economically and socially sound. Is such as approach possible?

Three imperatives

We suggest approaching the goal of integration by building on the concept of sustainable development, defined in terms of opposition to the three categories discussed above. Our central thesis is that it is critically important that all three prime systems move in the direction of sustainability and that these movements reinforce each other.

We suggest redefining sustainable development as the reconciliation of three imperatives:

+ The ecological imperative to remain within planetary biophysical carrying capacity;

+ The economic imperative to ensure and maintain adequate material living standards for all people; and

+ The social imperative to provide social structures, including systems of governance, that effectively propagate and sustain the values that people wish to live by.

In other words, sustainable development is for us inherently a normative concept, with each of these three imperatives being an ethical statement regarding its respective prime system. All three imperatives are more value laden than objective.

This is most evident in the case of the social imperative, which has both political and cultural components. The political component has to do with systems of governance that are feasible in the sense of being acceptable to citizenry and perceived as giving rise to a collective sense of well-being. The related cultural component has to do with the preservation and enhancement of social structures, including traditional cultures of various kinds. Together these two components are expected to foster a sense of community, an indispensable requirement for sustainable social well-being.

The economic imperative also implies a value judgment regarding the levels of nutrition, shelter, and material well-being that may be considered adequate and how these levels can best be reached (that is the meaning of *ensure and maintain*). One needs to consider both basic human needs and economic aspirations beyond simple subsistence requirements. Different countries have taken different approaches to satisfying the basic human needs of their populations;

clearly, the degree of need differs greatly between and within countries. Moreover, many different approaches are available to provide opportunities for people with larger economic aspirations. It is crucial that any attempt to promote sustainable development take into consideration the adequacy of material living standards, people's aspirations for higher living standards, and the methods for such improvements.

The ecological imperative at first sight appears more objective, but defining carrying capacity in terms that are not value laden is very difficult. The biomass productivity and mass–energy flux of largely untouched natural forests and prairie ecosystems may be substantially higher than those of the pastures or agricultural systems that replace them. But natural ecosystems can support only small human populations of hunter–gatherers, whereas pastures and agricultural systems can sustainably support far more people. Nevertheless, few people would argue that all natural ecosystems should be replaced by human settlements, industry, or intensively managed agriculture; at least some natural systems should be maintained to leave options for future generations, for ethical and spiritual reasons, and to maintain biodiversity for biotechnology.

Carrying capacity can perhaps best be defined as the maximum sustainable load, measured in terms of mass–energy flux of a desired kind, that can be extracted from or absorbed by a given biological system. This emphasizes the value-laden nature of the determination of what is desired, as well as emphasizing the biophysical limitations on achieving those desires (Dale et al. 1995).

Each of the three imperatives, therefore, appears to involve both objective and value-laden (ethical) components. The definitions suggested above need to be tested more widely, especially in cultures, societies, and economies other than those found in Canada. It may be possible to develop consensus on common global definitions of the three imperatives, at the same time leaving room for regional, national, and perhaps cultural and local variants of these definitions.

Some would argue that one imperative is more fundamental than the others: for example, that social justice is more important than economic well-being or that the objective of remaining within planetary carrying capacity is more basic than any other. In our view, just as the three prime systems are regarded as coequal, in that none has primacy over the others, so none of the three imperatives should be given special priority. Satisfying any one imperative without also satisfying the others would be unacceptable for at least three reasons.

First, each imperative is independently crucial. To ignore the social imperative is to risk serious dislocation in social structure and governance and to deny spiritual, philosophical, and societal needs. Humanity does not live by bread alone. To ignore the economic imperative is to accept that two-fifths of humanity will continue to live in absolute poverty. Finally, to ignore the ecological imperative is to invite ecological loss and disruption that will result in human health problems, massive effects on both economic well-being and social justice, and impoverishment of the planetary heritage.

Second, each imperative is urgent. We cannot afford to wait in any of these three areas. The scope and scale of the problems in each of these systems are such that we have no time left to stop and decide which is most important to deal with first.

Third, the three imperatives are interconnected. The extent of interaction among the three systems suggests that any attempt to address the issues of one system in isolation not only runs the risk of intensifying problems in the others but may also give rise to feedback effects from the other systems that would overwhelm the beneficial effects of the first intervention. For example, attempts to address environmental problems in isolation may involve policies that exacerbate economic or social problems, thus giving rise to further environmental stress. Consider wildlife-conservation programs in developing countries that displace people from a protected habitat, thus causing them to degrade the often marginal lands to which they are relocated (Guha 1989). Similar problems can occur with other economic and social systems; for example, economic policies in industrialized countries that create social tensions can in turn affect these countries' ability to compete internationally.

Because of the essential interconnectedness of the three prime systems and the urgency of the problems associated with each, addressing any one of the three imperatives in isolation virtually guarantees failure. Nevertheless, this is what policymakers commonly do. We argue for a more integrated approach, one that explicitly addresses all three dimensions of the overall problem. Truly sustainable development must reconcile all three imperatives. How can this be achieved?

Uncoupling economic growth from its environmental impacts

The linkage among the three prime systems is widely understood in terms of the interconnections between economic and social policies. The impact of taxation, access to credit, and investment patterns on employment, poverty, and child health, for example, are well recognized, if not always fully understood. The effects on the economy of social institutions and events such as education, disease, and migration are similarly recognized. More and more, however, as anthropogenic stresses on the biosphere have increased, socioeconomic linkages have become complicated by impacts from and on the ecological system, to the extent that environmental factors now often confound economic and social policies.

For example, irrigation and flood control have become dominant political and economic factors in Bangladesh and elsewhere. The economies and social structures of Sahelian countries are heavily influenced by desertification. Also, the long-term environmental problems of nuclear power (combined with its failure to win social acceptance) have proven a major brake on its use.

Although these interactions usually bring ecological and economic imperatives into conflict, these imperatives can be made compatible and even mutually reinforcing, at least in open, highly industrialized economies. What needs to be done in one arena may be complementary to and supportive of the objectives in another. To see this, it is helpful to examine the ways economic and ecological imperatives are discussed in some industrialized countries.

On the economic side, it is increasingly argued that in response to global economic integration, the high-wage, resource-based industrialized countries, such as Canada, must develop an economy based on goods and services with higher information content (RCEUDPC 1985). Such an economy is required if Canada is to compete in an increasingly integrated and competitive global market, characterized by very mobile capital and investment flows and decreasing barriers to such movements. Moreover, if the economic and social needs of developing countries, economies in transition, and rapidly industrializing economies are to be met, trade and flows of investment capital must be greatly expanded. For the poorest countries, such economic activity is needed merely to maintain their current, often inadequate, growth rates; moreover, industrialized countries increasingly depend on export revenues derived from trade with the rest of the world.

On the ecological side, there are calls for strategies and policies to encourage industry to develop industrial processes that are inherently more benign and to introduce production and consumption patterns that reduce the flow of matter and energy per unit of economic activity. Such "dematerialization" strategies would reduce environmental impacts in industrialized countries and lead to new technologies for export, thus reducing environmental impacts in other countries as well. Some evidence shows that much of the technology currently exported from industrialized to developing countries is not state-of-the-art in terms of environmental impact and matter–energy input but is cheaper technology that is already obsolescent in the home country. This represents, not dematerialization, but increased materialization in these economies.

The original argument for reducing the throughput of matter and energy per unit of economic activity was made by Daly in 1980, in the context of arguments in favour of a steady-state economy (see Daly 1991). Since then, the arguments have been extended to encompass dematerialization, inherently benign industrial processes, ecoefficiency, and "green" technologies (Williams et al. 1987; Kanoh 1992; Robins 1993; BCSD 1994).

The economic and ecological arguments presented in this section raise an interesting question: To what extent are the measures required for living within our ecological carrying capacity compatible with, or even necessary to, the measures required to meet the challenge of economic restructuring driven by global economic integration?

Must economic and environmental agendas conflict?

Society and business may have a strategic opportunity to go beyond thinking that environmental and economic agendas necessarily conflict, that economic activity undermines sustainability, and that ecological sustainability constrains economic activity. The ecological and economic imperatives could reinforce each other, if they are interpreted and acted on imaginatively.

If, for ecological reasons, Canada and other industrialized countries need to dematerialize their economies by uncoupling human well-being from the throughput of matter and energy in society, they will have to develop more eco-efficient technologies (that is, substitute knowledge and efficient design for waste-ful, "brute-force" technologies). Conversely, if, for economic reasons, Canada needs high value-added, information-intensive industries that maintain interna-tional competitiveness and generate high standards of living, it will have to develop industries characterized by their innovation and information content and by principles of advanced design and management. In either case, international trade becomes a crucial factor in achieving or undermining such connections.

In other words, the measures needed to achieve ecological sustainability may, at least in a country like Canada, also be the measures needed to maintain a competitive niche in the global economy. Substituting information for the throughput of matter and energy would allow Canada to capitalize on its edu-cated work force as a source of comparative advantage and, at the same time, reduce the country's environmental damage per unit of economic activity (the production and processing of information can involve much lower use of phys-ical resources). Achieving such simultaneous gains in productivity and environ-mental impact through dematerialization of the economy is the basic premise of industrial ecology (Allenby 1992; Frosch and Gallapoulous 1992; Tibbs 1992; Hawken 1993; Ekins and Jacobs 1994; Young and Sachs 1994; Cote and Plunkett 1996).

Of course, posing the issue in this fashion raises important questions. To what degree would the growth in economic activity resulting from such eco-nomic restructuring increase the ecological impacts? In other words, would growth in activity levels more than offset increases in efficiency? Would changes in lifestyle and consumption patterns, such as substituting social-support processes for material consumption, be a more useful approach to staying within ecological carrying capacity?[5]

---

[5] Dematerialization and a substantial increase in knowledge-based industry in industrialized economies may provide a powerful economic incentive for a highly educated, healthy, and socially stable work force, which may tend to satisfy some components of the social imperative. The inter-national implications of this are less clear. Knowledge-based industries in Singapore, for example, rely heavily on immigrant, unskilled female labour. This, however, may be a transitory phenome-non; globalization of the economy may indeed temporarily shift unskilled jobs to where there is a pool of cheap, unskilled labour while the achievement of higher education and other social goals fol-lows dematerialization.

Another important set of questions relates to the distribution and equity issues raised by either type of strategy and the ethical and practical issues raised by the large and growing fraction of the globe's human population existing in a state of extreme poverty. We will return to these questions later.

In any case, the increasing importance of international trade and investment makes it necessary to consider a sustainable-development approach of the type we suggest. This strategy would explore the prospects for enhancing Canada's international competitiveness through economic and trade policies based on the development of environmentally benign, clean, green technologies and technological systems with minimized matter–energy throughput (which may in the long run be as important as limiting pollution). This goes beyond the development and export of end-of-pipe waste-treatment plants or emission-control technologies and implies the need to develop new forms of industrial processes that are inherently benign. The focus is on the design of such processes, a focus that is more fundamental and also more supportive of an economic culture based on new knowledge and innovation. This perspective implies a significant emphasis on systems of technological innovation geared to competitive, environmentally benign industrial processes. This in turn is likely to require forms of social and political organization that promote such innovation and provide some confidence in the belief that the rewards of such processes will be shared equitably.

A dematerialization strategy would involve a progressive uncoupling of economic activity from the throughput of matter and energy in a society. Such dematerialization could result in a "factor 10" economy: one that, as a result of measures that drive a "policy wedge" between economic activity and matter–energy throughput, uses only 10% of the matter and energy used today per unit of economic activity (Schmidt-Bleek 1992a, b, 1993; von Weizsacker 1994; see also the reservations expressed in Rees 1995).

The dematerialization policy wedge

The concept of a policy wedge between economic activity and matter–energy throughput is shown in Figure 1. Traditionally, rising standards of living (economic growth) involve growing consumption of goods and services, coupled with greater consumption of energy and materials, leading to higher environmental impacts. In this scenario, economic and environmental goals are in conflict, with one achievable only at the expense of the other. The dematerialization strategy in Figure 1 would involve uncoupling economic growth from environmental impact by developing more environmentally benign technologies. This drives a policy wedge between the consumption of goods and services and matter–energy throughput, thereby helping to reconcile economic and environmental goals.

Although the argument has so far been expressed in terms of industrialized countries like Canada, similar arguments may apply to many developing and rapidly industrializing nations: matter–energy throughput could in many

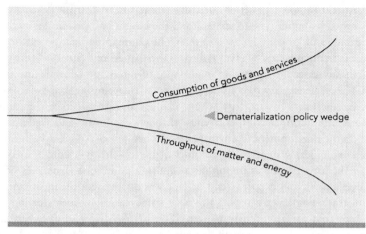

Time

Figure 1. Dematerializing the economy. The first policy wedge, dematerialization, uncouples economic activity from matter–energy throughput. Because matter–energy throughput heavily influences environmental impact, dematerialization can reconcile environmental and economic goals. Source: Rees (1995), based on Pearce (1994).

ways be uncoupled from economic growth. The development of technological "leap-frogging" strategies in developing countries, whereby intermediate "dirty" stages of industrialization are bypassed as these countries move directly to more advanced and environmentally benign technologies, may offer significant hope for improvement in terms of ecological damage, quality of life, and economic sustainability (Berrah 1983; Goldemberg 1992). Also, as per capita income in such countries increases, it can be expected that demand for improved environmental quality will grow rapidly (World Bank 1992). Moreover, if a dematerialization approach is to improve the international competitiveness of industrialized countries, there must be a growing domestic and international demand for the goods and services produced through such a strategy. In this sense, the success of such a strategy depends on the existence of an international market for such products, which should in turn reduce environmental impacts directly and stimulate the development of more environmentally benign production processes throughout the world.

The issue is of course more complicated. The net environmental and economic impacts on developing countries of industrialized countries' pursuit of a dematerialization strategy depend on a whole host of factors, including the types of goods traded, the terms of trade, and the indigenous technological capability of the developing countries. Most developing countries are far less "marketized" than industrialized countries and have much larger subsistence and informal economies. We do not know how much dematerialization could reduce total environmental impacts in societies in which basic needs are still widely unsatisfied and in which considerable sustained economic growth would be needed to achieve this.

Another important issue is gauging the macroeconomic effects of dematerialization strategies. Sanders (1992) and Simpson (1995) argued that insofar as increases in energy efficiency and reductions in environmental impacts contribute to reductions in input costs, they will induce extra economic growth in affected sectors, thus reducing or offsetting any efficiency or pollution-reduction gains. The existence, nature, and size of such rebound effects are critical challenges for the dematerialization strategy.

These uncertainties underline the need to investigate whether a complementary set of dematerialization strategies could be devised for the major economic regions of the world, strategies that would reinforce each other through international trade and improve human well-being and ecological conditions across a range of societies, cultures, and economies. Such dematerialization strategies go beyond the current rhetoric (let alone the practice) of the *post hoc* integration of environmental concerns into economic decision-making, as dematerialization strategies can be mutually reinforcing. Thus, they may offer significant potential for integrating the ecological and economic imperatives.

Resocialization: the second policy wedge

We also need to consider the social imperative, and here a number of concerns emerge, as the reconciliation of ecological and economic imperatives through dematerialization, suggested above, involves complex interactions with the social imperative. For example, growing global economic integration seems to produce growing social and cultural fragmentation. Would dematerialization strategies reinforce such tendencies? Certainly it is possible to imagine a version of this approach that would, at least in the short term, simply reinforce globalization's tendency to increase income disparities. In this scenario, countries and sectors within countries that succeed in pursuing the reconciliation outlined above would continue to prosper at the expense of everyone else. In particular, highly educated elites in most countries would be able to reap the benefits of a more integrated and interconnected global economy and would in so doing be increasingly able to buy whatever level of environmental quality and social security they desire (Schrecker 1998).

The problem is not so much that the rest of the world would fail to benefit from this process, as the successful pursuit of dematerialization strategies is likely to foster the emergence of prosperous markets for more environmentally benign technology and processes; it is unclear, however, whether the rest of the world would necessarily benefit from this process to the degree and in the time frame required for these strategies to be politically feasible. If dematerialization strategies themselves do not incorporate measures to address the social imperative directly, then they are likely to fail because the degree of social coherence required for such strategies to work will be unavailable. In this sense, a trickle-down approach to the social aspects of dematerialization would be morally questionable and probably unfeasible.

A look at the issue of employment illustrates these complexities. Although dematerialization strategies will, by definition, reduce environmental impact per unit of economic activity, the labour and employment implications of this approach are much less clear. The information economy may substitute information for matter and energy, which will reduce environmental impacts, but it might also substitute information for labour, which will reduce employment. To a significant degree, the question turns on whether information and labour are substitutes or complements. If information and labour are substitutes, as information and matter–energy are, then using more information will reduce the use of labour along with the use of matter and energy. If information and labour are complements, then using more information will also use more labour. In the latter case, we would have the best of both worlds: more information, more jobs, and reduced environmental impact.

Of course, the reality is rather more complex. Whether information is a substitute or a complement for labour or for matter–energy depends on the circumstances, the technologies involved, and the relationship among all three inputs. However, although there is a reasonable argument for substituting at least some information for at least some matter and energy (the basis of the dematerialization argument), it is much less clear that employment levels will be maintained or increased by such processes.

In Canada, for example, the effect of new technology in the British Columbia forest industry in recent decades has been to reduce labour per unit of production, with progressively fewer people employed to produce progressively more timber. A similar phenomenon has been widely observed in developing countries, where the agricultural work force per hectare has been reduced as cash crops have replaced subsistence farming. Considerable thought needs to be given to the employment implications of a strategy of economic dematerialization and the capacity of social systems to accommodate such changes.

These considerations suggest that it is crucial to begin thinking about ways to integrate the social imperative into our response to economic and ecological problems. Even if the dematerialization approach reduces negative environmental impacts and improves aggregate economic conditions, it is unlikely to be politically feasible unless it also reduces social tensions. Problems of governance and culture may be the real limiting variables in achieving sustainable development, not least because resolving these problems is necessary for facilitating the required changes in the economic and ecological spheres.

One way to address this issue may be to start by considering the obvious fact that human well-being and economic activity, although connected, are not synonymous. Economic indicators, such as gross domestic product (GDP), are incomplete and sometimes misleading measures of well-being, for a number of reasons:

+ Changes in the composition of GDP may have major impacts on human well-being that are invisible to an analysis that considers only the level of GDP;

+ Human well-being may not increase linearly with consumption, so consumption changes do not necessarily indicate commensurate changes in well-being;

+ Changes in the level of GDP fail to account for the relationship between the distribution of income and overall well-being;

+ GDP takes into account only activities in the formal economy and leaves out a host of informal economic activities and intangible, or nonmarket, effects; and

+ Environmental degradation reduces well-being but does not result in a corresponding reduction in GDP.

Various attempts have been made to construct alternative measures of social well-being (Daly and Cobb 1994; Gustavson and Lonergan 1994). For our purposes, the distinction between economic activity and human well-being may provide a focus for a second policy wedge that links the economic and the social. What mostly seems to be missing in using measures of economic activity as a proxy for human well-being is precisely those sociopolitical and cultural factors that have to do with a sense of community, social coherence, and collective well-being. From this perspective, fostering economic growth and increasing environmental quality may be necessary but radically insufficient. In particular, we may need to explicitly recognize that some aspects of human well-being cannot be reduced to economic activity but need to be fostered on their own. This approach has been used by the United Nations Development Programme to create a human-development index that incorporates GDP but also includes levels of education and health (UNDP 1994).

Although the almost universal tendency in modern political decision-making is to maximize economic growth, with the expectation that such growth will lead to increases in human well-being, we suggest that there is a need to develop policies to increase human well-being per unit of economic activity (for example, per dollar of GDP). In other words, social strategies that uncouple human well-being from economic activity are needed to complement dematerialization strategies that uncouple matter–energy throughput from economic growth. The effect of such "resocialization," shown in Figure 2, would be to drive a policy wedge between human well-being and the consumption of goods and services, allowing the former to rise while the latter falls.

What might some of these resocialization strategies look like? One possibility in industrialized countries is a shift to having a society with less participation in the formal economy, for example, through a shorter average work week, coupled with greater leisure and more participation in activities of the informal economy, such as community service of various kinds, child care, coaching, and managing recreational sports. The World Alliance for Citizen Participation, for example, recently discussed strengthening civil society through increased emphasis on the informal economy (CIVICUS 1994). Also, the

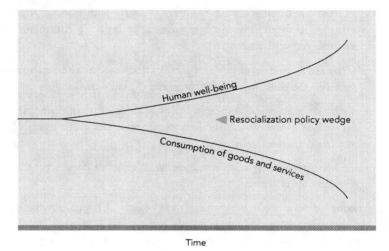

Time

Figure 2. Resocializing society. The second policy wedge, resocialization, uncouples human well-being from the consumption of goods and services. Resocialization can reconcile economic and social goals.

Commission on Global Governance has studied general issues of governance, security, and economic interdependence (CGG 1995), and Etzioni (1993) examined communitarian approaches to these issues in a North American context, emphasizing the re-emergence of community and informal-economy activities.

To make resocialization possible, we might consider making investments in social capital that parallel our investments in natural capital. Dobell (1994) proposed a guaranteed annual income scheme, to be paid for by revenues generated through an ecological tax reform. The intended effect was to uncouple work from income and to invest in the social infrastructure to allow people more time for filling socially useful roles, as well as to free up labour markets.

A shift in emphasis from the formal economy relies on people being willing to take part of their income in the form of increased leisure and unsalaried activities. A trend toward this is found in a number of industrialized countries where people are actively exploring job-sharing arrangements of various kinds, although this is typically driven by unemployment problems, rather than a desire for shorter work weeks and lower incomes.[6] However, if such a shift in economic activities is imposed on unwilling citizens while other socioeconomic groups maintain full employment and high incomes, further social unsustainability might result. The need for resocialization policies in this situation becomes apparent.

---

[6] Our attempts to model sustainable-society scenarios for Canada revealed that a reduced average work week (that is, implicit increases in job-sharing) is one of the few ways to maintain something close to full employment in an economy with lower consumption levels (Robinson 1998).

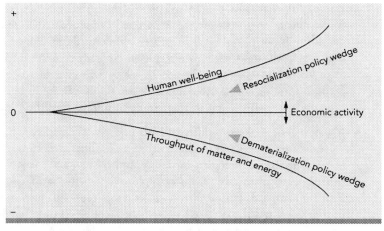

Time

Figure 3. Reconciling human well-being and ecological carrying capacity. The combination of dematerialization and socialization can lead to win–win–win futures, in which human well-being rises and matter–energy throughput (and thus environmental impact) falls. Economic activity (consumption of goods and services) could rise, fall, or remain steady; it might well rise in poorer societies and fall in richer ones.

Combining the policy wedges

Figure 3 shows the possibilities when the first policy wedge, dematerialization, is combined with this second policy wedge, resocialization. This represents a significant elaboration of the dematerialization strategy outlined in Figure 1. In Figure 3, economic activity (the middle lines) is progressively uncoupled from matter–energy throughput (bottom line), through dematerialization strategies, and human well-being is progressively uncoupled from economic activity (top line), through resocialization strategies. Over time, throughput of matter–energy falls and so, therefore, do impacts on ecological carrying capacity. Human well-being rises, and economic activity (middle lines) stays somewhere in between, rising significantly in many parts of the developing world and among disadvantaged communities in the developed world but increasing moderately, remaining the same, or even falling in parts of the industrialized world.

This approach explicitly recognizes the importance and influence of the social dimension, in contrast to the pure-dematerialization approach illustrated in Figure 1. The approach summarized in Figure 3 offers a more hopeful and, we believe, more realistic prospect, for three reasons:

+ It recognizes that without addressing questions of governance and culture, we are unlikely to develop and successfully apply dematerialization strategies. In other words, it explicitly distinguishes the three dimensions of the problem.

✦ It recognizes that social choices and policies themselves can help reduce environmental impacts. The whole burden of improved environmental quality is not placed on the dematerialization strategy alone but is shared by both strategies. This allows us to explicitly consider the issue of overconsumption (Durning 1992), which was introduced to the political agenda at the Earth Summit in Rio (UNCED 1992) but had been discussed without much real resolution since at least the 1970s.

✦ It allows policymakers in different parts of the world to take substantially different policy approaches, according to differences in poverty, material needs, technological capability, and cultural and social goals. In some areas of the world, for example, where better satisfaction of basic human needs is a priority, it is likely that economic growth will be a priority, and the emphasis should therefore be on dematerialization and reducing environmental damage. In other areas, where high levels of material consumption are already typical, the focus may be more on resocialization policies that uncouple human well-being from economic activity.

Both dematerialization and resocialization are likely to depend heavily on there being adequate institutions, both state-based ones and ones within civil society; there is great variation, however, in the strength of such institutions, especially in developing and former communist countries. The overall aim of the combined strategy is to increase human well-being and to decrease environmental damage, and this can be accomplished in different ways in different areas of the world. In this scenario, economic activity is not an end in itself, but merely a means to ecological and social goals.[7]

This dual role for resocialization policies — making dematerialization strategies possible and directly reducing environmental impact — emphasizes the importance of the social imperative. Fortunately, some encouraging examples can be given of such approaches. These include greater recognition and use of the informal economy, embracing the concept of "sustainable livelihoods," rather than salaried employment; more community self-management, such as small-scale and community credit schemes; and lower cost, less impersonal ways of providing goods and services, such as home health care and the use of paramedics, rather than hospitalization, or small, intensive family farms and urban

---

[7] Indeed, the economic imperative may only be a transitory phenomenon, which will ultimately disappear when living standards uniformly reach an acceptable level. This is one reason why this chapter defines the economic imperative as the need to ensure and maintain adequate material standards of living for all people. Given the grossly inadequate material living standards of about two-thirds of humanity at present (including not only the majorities in many developing countries but minorities in richer nations, such as the old, the sick, the unemployed, and the disabled, as well as many cultural, ethnic, and other disadvantaged groups), the economic imperative will clearly be with us for some time.

vegetable gardens. However, an effective and more widespread uncoupling of economic activity from human well-being would address more intractable problems, lying mainly in the social system: the consumer-society ethic, which equates human well-being with material consumption, and the degree to which global economic integration, as currently practiced, tends to decrease the sense of community, social coherence, and governability.

This integrated policy direction implies an optimistic scenario: a win–win–win solution in which all three imperatives are satisfied. However, this cannot be guaranteed to happen in any given country, let alone in all countries. Indeed, the current tendency is to concentrate on the economic imperative, combined with a *post hoc* attempt to reconcile this with the ecological imperative, while largely ignoring the social imperative and questions of North–South and intracountry equity.

Sustainable development will not be achieved through technical fixes alone. A strong social dimension has to do not only with political issues but also with values, lifestyles, social organization, and individual and collective behaviour. Both dematerialization and resocialization are needed.

Policy implications of a more integrated approach

In virtually all states and at the international level, public policy is determined sectorally within each prime system by separate ministries or agencies pursuing relatively narrow and sometimes conflicting goals. This is most clearly shown in the case of economic policy, which is determined by finance, trade, and similar agencies, sometimes with attempts to later adjust these policies to address various ecological and social objectives.

A number of attempts to do better than this have been made globally, nationally, and locally. The two iterations of the World Conservation Strategy, the Brundtland report, and the 1992 Earth Summit were all endeavours to find consensus on common international goals and strategies. The Netherlands, Canada, and Norway are among states that have tried to do the same nationally. But none of these attempts have resulted in any successful strategy or practical policies to fully reconcile the three imperatives. This is perhaps because the processes for developing such broad national strategies have tended to be dominated by environmentalists and environmental agencies and lacked high-level participation from policymakers and managers dealing with social and economic systems.

Initiatives with a more restricted scope have arguably shown greater promise. In 1994, for example, Canada's International Institute for Sustainable Development convened a small international group of government representatives, consumers, diplomats, trade experts, and environmentalists, who developed and released the Winnipeg Principles for Trade and Sustainable

Development. These principles represent an attempt to establish agreed sustainability yardsticks for measuring trade policies. International trade, worth around 6 trillion US dollars a year, has a powerful influence on all three prime systems. By creating wealth, international trade can reduce poverty and help address environmental degradation; conversely, some of the economic activities that trade accelerates can damage the ecological system and decrease equity among and within societies.

The Winnipeg Principles have established substantial common ground among those concerned with managing all three prime systems. The seven principles, all of which must be respected if a trade policy or agreement is to contribute to sustainability, address efficiency and cost internalization; equity (within and between countries and generations); environmental integrity; subsidiarity (acting at the lowest practicable jurisdictional level); international cooperation; respect for ecological science and a precautionary approach that recognizes uncertainty; and both openness and accountability.

Another policy initiative that may show us how to reconcile the imperatives is the response of the health-care system in some developed countries to the AIDS pandemic. Initially in San Francisco and subsequently in many other cities and countries, the high cost of both public education and prolonged palliative care led the health-care system to subcontract significant components of these to voluntary groups. The World Health Organization has argued that this is often less expensive and more effective. For example, AIDS typically involves short acute episodes, which may require hospital treatment, interspersed with long periods when only minimal nursing and monitoring are needed. This disease pattern made it possible to develop the "buddy" system, with home care provided by largely untrained volunteers at a fraction of the cost of permanent hospitalization, which has major benefits in terms of personal independence and dignity. Also, the perceived need in nearly all countries (derived from the economic imperative) to limit or reduce public health-care expenditures is commonly thought to involve reductions in the standards of health care. But this AIDS example shows that this need not be the case: there are ways to both improve care and reduce costs (Mann et al. 1992).

Some promising initiatives in reconciling the imperatives have arisen because the economic imperative is influencing virtually all governments today to reduce the size and costs of bureaucracies. Because this appears to militate against more environmental regulation and control, interest is growing in the use of fiscal measures to achieve environmental goals. One recent study (IISD 1994) documented 23 European and North American cases in which taxes and subsidies are used more or less successfully to achieve environmental goals. A related approach, being explored in a number of industrialized countries, is to have environmental (or sustainability) audits of corporations and industries carried out annually by independent external specialists (analogous to financial auditors). This may reduce the costs and bureaucracy of government regulation while making industry more self-policing and publicly accountable.

It seems probable, at least at this stage, that relatively limited initiatives of these types may have a greater practical impact on public and private decision-making than attempts to develop more comprehensive global agreements. However, if such new approaches are to integrate the three dimensions of the problem, they must have a policy framework that recognizes the need for such an integration. We have two suggestions: subsidiarity and binding agreements; and a constrained market. They are not worked out in detail but provide a general direction for how this might be done (see also Robinson and Van Bers 1996; Robinson 1998).

Subsidiarity and binding agreements

On the sociopolitical side, increasing pressures for recognition of cultural and ethnic differences, combined with a call for more localized decision-making, suggest the need to further decentralize political power over cultural and the environmental matters. However, owing to the increasingly global nature of ecological and economic problems, decentralization by itself is an insufficient response. Instead, we propose a form of devolution of political power and management authority, combined with a much stronger common legal framework of environmental and cultural rights. The price for such devolution would be explicit acceptance of agreed cultural and environmental standards, such as protection of minority-language rights or legally binding water-quality standards.

This approach would require an innovative form of contract, whereby larger political units expressly delegate management power to smaller jurisdictions (subsidiarity) in return for legally binding agreements on basic environmental and cultural rights and responsibilities at both the individual and community levels. This would allow for more local political involvement in environmental management and cultural development at a spatial level that may be more in tune with immediate experience and thus more likely to be both socially acceptable and effective. Establishing binding agreements on basic environmental and cultural rights and responsibilities would require a process for defining such rights and responsibilities. This might be a process of public consultation at the national level (linked to international negotiations) to propose, debate, and determine environmental strategies, programs, goals, and targets. These targets would be broadly defined and implemented at as low a level of jurisdiction as possible. Pollution-control goals, for example, would be more likely to involve environmental-quality standards than emission standards or specific forms of emission technology. A social goal might involve an agreement that the average income of the top 20% of the population exceed that of the bottom 20% by no more than a certain factor, leaving it to individual nations or provinces to decide how to achieve this target (or by how much to exceed it).

The essential components of this subsidiarity approach are that common goals should be agreed in common; that management to achieve these goals would be devolved to as low a level of jurisdiction as possible; and that the lower jurisdiction would be formally accountable for achieving the targets. In many respects, this combination of subsidiarity and binding agreements can be summed up in the phrase "sovereignty association," a term made prominent in the Canadian debate over the political separation of Quebec from the rest of the country. A similar sense is also evoked in the well-known environmental slogan "think globally and act locally." Each expresses a need to combine local control and strong links to the larger context. Internationally, such an approach could develop logically from existing agreements, such as the United Nations Declaration on Human Rights or the Montréal Protocol on Substances That Deplete the Ozone Layer; nationally, many states have agreements similar to the US *Bill of Rights* or the Canadian *Charter of Rights and Freedoms.*

It can be argued that although this approach is conformable with the tradition of Western industrialized countries, it would be unlikely to appeal to governments of many former communist and developing countries, where individual and minority rights tend to be regarded as a luxury competing with economic growth and where greater emphasis is traditionally on the unitary state. Certainly, respect for individual, cultural, and other minority rights is likely to continue to be uneven, but global and regional free-trade agreements, which for most countries are a part of the economic imperative, are likely to continue to involve basic environmental and social standards, however limited and patchy they may be at first. Moreover, as suggested above, a more equitable distribution of social and economic goods may be necessary to create and maintain the social cohesion and stability needed for economic restructuring.

The constrained market

On the economic side, we suggest a constrained market, whereby extensive use is made of market-based instruments (MBIs), wherever feasible and efficient, but these are backed up by powerful, politically determined constraints and nonmarket implementation measures. Like virtually all existing economic regimes, a constrained market represents a middle ground between pure reliance on the market and pure reliance on government control. The difference is that this approach would define that middle ground partly in terms of sustainability goals. Constraints on the market would be of two general types: boundary, or external, constraints that set limits within which the markets may operate; and target, or internal, constraints that alter the market value of goods, services, and resources.

The major mechanisms for deciding such constraints would be the environmental target-setting and binding-agreement processes described above. Those processes would produce a set of targets, based partly on an analysis of the environmental implications and economic feasibility of attainment. This analysis would then be used to determine which standards are overriding and would simply be imposed, regardless of economic conditions (boundary constraints) and which ones were to be subject to constraints within the market system (target constraints). Criteria for deciding the type of constraint to be used might include the scale of the environmental impact, whether it is reversible, uncertainty about impacts, and the economic cost of imposing a ban or other boundary constraint.

If a boundary constraint is indicated, such as zero discharge of particular toxic and persistent chemicals, then this would be applied through passage and implementation of standards and regulations of the traditional kind. If, however, analysis determined that target constraints would suffice, then this would lead to the design and implementation of appropriate MBIs, such as taxes, incentives, or tradeable permits. In this case, the target-setting analysis could be used to estimate how incentives would have to be changed so that the desirable standards and targets can be reached. Such changes would then be instituted in the form of taxes and charges on (or possibly subsidies for) goods, services, and resources, including such mechanisms as effluent charges or sale of pollution permits.

Markets never work perfectly. Even if target constraints are used, there would likely be a need for nonmarket programs to supplement the MBIs. Such measures could, first, supplement the modified market signals resulting from the imposition of target constraints and, second, substitute for such signals altogether in the case of boundary constraints. These implementation processes would involve direct government intervention in the marketplace and could be based on the lessons learned from past environmental and energy policies. The measures might include direct instalment of environmentally friendly equipment, accelerated replacement of capital through subsidies, and involvement of local community groups in program design and delivery.

The development and implementation of target constraints would allow the marketplace considerable freedom to choose the least-cost method of achieving environmental targets. This would amount to an internalization of environmental costs and benefits, based on a political valuation of environmental costs and benefits not captured through normal market processes, such as the "bequest value" to future generations and some of the "option value," or opportunity cost, of environmental amenities. The MBIs used to implement target constraints could also generate government revenue. The overall effect would be to reap the benefits of economic efficiency without succumbing to the short-term focus and lack of goals implicit in traditional market approaches or to the rigidity and inefficiency of purely regulation-based approaches.

Dematerialization and resocialization are two key strategies for reconciling the three imperatives, and subsidiarity and binding agreements and the constrained market are two possible approaches to the policy implementation of such strategies. Although global consensus on the three imperatives and on complementary national strategies for addressing them is clearly needed, policy implementation will vary according to national and local circumstances, rather than being internationally homogeneous.

The need for better analytical frameworks

Our current state of knowledge about each of the three prime systems is probably too insufficient and certainly too inadequate, given their interactions, to allow us to manage any of them effectively in isolation. Scientific and other scholarly studies, both theoretical and policy oriented, are still sectorally fragmented by academic and bureaucratic boundaries, with major differences in vocabulary, concepts, and unstated assumptions. The recognition that the ecological, economic, and social spheres are largely coterminous, interacting, complex, self-organizing systems, each coequal in primacy, as argued above, may provide some common ground. So might the application of a common systems-analysis approach to managing all three prime systems.

We developed this framework, based on the three prime systems and the three imperatives, primarily in response to conditions in Canada and in other industrialized countries. Careful consideration and adaptation are needed to apply this framework in rapidly industrializing countries, in countries in transition from command to market economies, and in other developing regions of the world. For example, in working to reconcile the three imperatives in other societies, where does one find similar policy wedges? Are the various policy wedges in different societies compatible?

Arguably, the research needed the most urgently is the development of a common analytical framework that functions equally well in all three prime systems and among their interactions: a tool to measure the success of strategies and policies from the local to the international levels and to determine the extent to which they contribute to developing more-sustainable societies and to reconciling the three imperatives.

The limitations of existing analytical tools

Many analytical tools have been developed to address aspects of these issues. Perhaps the most commonly used are environmental-impact assessment (EIA); social-impact assessment (SIA), which is often combined in various ways with EIA; cost–benefit analysis (and its variants, such as cost-effectiveness analysis); and, more recently, integrated assessment (IA). Bailey et al. (1995) recently reviewed cost–benefit analysis and other analytical approaches in the context of an IA of environmental change (see also Cohen 1996; Rothman and Robinson 1997; IPCC 1998).

With the exception of IA, such analyses are typically focused on only one aspect of one of the three systems. For example, EIA is often combined with SIA and with economic analysis in various ways, but the central focus is on ecological impacts and their direct consequences. Similarly, cost–benefit analysis typically addresses only the direct and indirect economic costs and benefits of a project or policy. Although such analyses may provide very useful information for decision-makers, they clearly do not address the full range of factors.

When we turn to IA approaches, the scope is broader but still inadequate from the point of view of integration. For example, to date, the usual attempt to develop IA of options to mitigate climate change is to apply a comprehensive cost–benefit framework to the analysis of alternative climate policies, including an explicit attempt to estimate the costs associated with some environmental and social impacts. This is done to provide some common dollar metric with which to weigh the positive and negative impacts of different policies over time. One should applaud the attempt to systematically compare both the positive and the negative impacts of various options or scenarios. However, expressing these impacts in purely monetary terms raises many concerns:

+ Economic activity is not a direct measure of well-being;

+ Many intangible and effectively unquantifiable effects need to be considered;

+ Serious methodological problems plague such indirect measurement techniques as contingent valuation;

+ Fundamental disputes about whether markets should be assumed to operate efficiently and about how issues of equity and discounting should be addressed are waged in the literature; and

+ Grave difficulties hamper the application of existing macroeconomic models to economies in transition or those of developing countries.

Moreover, purely economic factors are clearly only one input to complex political and business decision-making processes, which often seem resistant to the results of economic analysis. (In this sense, economic analysts often seem to mimic the political naivete of natural scientists in the environmental field, who sometimes act as if they believed that the results of their analysis should dictate policy decisions.) The point here is not that decisions are in practice made for irrational, noneconomic reasons but that the scope of purely economic analyses may be too narrow to encompass crucial dimensions of the problems of interest to decision-makers.

More recently, in response to dissatisfaction with our ability to meaning-fully express very different kinds of effects in monetary terms, a growing interest has emerged in multiattribute forms of analysis that make no attempt to express all values in terms of dollars. Such approaches avoid the mistake of trying to force all issues into one metric, and they allow users of the analysis to apply their own weighting schemes and priorities to the results. But these analyses do not go far enough in addressing social and economic issues, as they are still too focused on the direct, quantifiable impacts.

These considerations suggest the need for new analytical frameworks to supplement the more specific and detailed, but narrower ones and provide a much broader and more integrated view of the problems.

Three suggestions for an improved analytical framework

To satisfy the needs of integration, any new analytical framework must address two main concerns. First, the substantive scope of analysis needs to be broader, embracing social and economic issues, such as migration, equity, income disparities, governance, cultural diversity, international trade and competitiveness, and investment, as well as biophysical issues, like species loss, pollution, resource scarcity, and the maintenance of ecological life-support systems. Second, the methodological scope of analysis needs to be wider, going beyond impact assessment to a much more integrated assessment of such things as alternative scenarios, development paths, and the implications of lifestyle changes and systems of governance.

One fruitful approach may be to recognize the complex, self-organizing nature of the ecological, economic, and social systems and the interactions among them. This means a strong focus on interactive effects, second- and higher-order consequences, nonlinearity, thresholds, and changes in state, diversity, and resilience. In turn, this suggests the need to develop integrated modeling and scenario analysis to allow us to treat each of these systems explicitly in terms of alternative development paths, rather than in terms of probable futures (Robinson 1988, 1991). This also means that economic, ecological, and social factors should be considered at the system level, not just in terms of direct impacts. Of interest are not simply the direct economic costs of various measures or effects, but their implications in terms of, say, trade or competitiveness. Social and ecological effects go beyond their direct impacts, like job or species losses, and encompass broader changes at the community or ecosystem scale.

Moreover, we need to explicitly consider interactions among these systems. We might start from the observation that each of the prime systems generally responds to stress, first, by absorbing it, through a variety of buffering mechanisms; second, by small, incremental, and evolutionary changes; and third, by major flips, or phase changes, which normally appear to be sudden, unidirectional, and irreversible. The direction, quality, and extent of such major phase changes may be unexpected, counterintuitive, and unpredictable.

A second dimension of more integrated analytical frameworks is the need to explicitly recognize the qualitative aspects of social and economic systems. Many assessment methods, based as they are on purely quantitative methods, are blind to such crucial issues as power, control, sense of community, trust, nationalism, and cultural identity. These issues are typically addressed in the humanities or the more qualitative parts of the social sciences (Robinson and Timmerman 1993). Any IA designed to comprehensively survey the implications of alternative patterns of developments or sets of policies must pay significant attention to such qualitative dimensions.

A third suggestion is to use and further develop approaches to analysis that are inherently interdisciplinary and policy relevant and involve the user community, not just as the audience for the published results, but as partners in designing and sometimes in undertaking the research.[8] Such an orientation changes the focus and design of the research and the nature of the interaction between analysts and the user community in ways that usually produce a more integrated picture of the issues.

The net effect of these suggestions would be the development of analytical frameworks based on constructing alternative scenarios and describing key interactions within and among the ecological, economic, and social systems. The goal of these frameworks would be to compare the relative effects of these scenarios in terms meaningful to all participants, using state-of-the-art understanding and techniques. Scenarios would be both qualitative (for example, story telling) and quantitative (for example, formal modeling) and would focus on issues such as choice, constraints, uncertainty, and surprise.

The type of analysis we are recommending would not of course replace more traditional approaches to IA, cost–benefit analysis, EIA, or SIA. Such conventional tools, which can provide a much more detailed examination of certain economic and biophysical dimensions of the problems, will continue to be used. Because the approach to IA we propose is necessarily highly aggregated and simplified, this type of analysis cannot do more than supplement existing methods. Nevertheless, this supplementary role is very important in addressing the kinds of issues involved in integration.

We are suggesting a form of IA going well beyond those typically found, for example, in the climate-change literature (Rothman and Robinson 1997). Our suggestions pose some formidable practical, methodological, and theoretical obstacles, including problems of data availability, problems of aggregation and integration, serious communication problems among and between different disciplines, lack of clear conceptual frameworks and theories, and uncertain prospects for success, either analytically or in terms of policy relevance. From one point of view, these problems may seem daunting. From another, they are challenges merely reflecting the need to change what is typically done.

---

[8] Such an approach, which is embedded in the mandate of the Sustainable Development Research Institute, has strong parallels with the concepts of "civic science," "vernacular science," and "post-normal science," as suggested by O'Riordan (1991, 1994) and Funtowicz and Ravetz (1993) (see also Wynne 1992; Shackley and Wynne 1995).

Chapter 2

Learning and Policy Integration

A.K. Bernard and G. Armstrong

❖ ◆ ❖

The setting

The truth is that all-out economic growth can no longer be viewed as the ideal way of reconciling material progress with equity, respect for the human condition and respect for the natural assets that we have a duty to hand on in good condition ... . We have by no means grasped all the implications of this as regards both the ends and means of sustainable development and new forms of ... co-operation.

Delors (1996, p. 13)

Integrating policy to simultaneously address social, economic, and environmental goals can be neither a simply conceived nor easily implemented solution to complex problems. Rather, it involves "progressively introducing and reinforcing a willingness to consider environmental effects ... [not] in a planned linear fashion but ... incrementally in response to specific demands or opportunities" (DePape, this volume, p. 113). Although it is clear that the state cannot act alone in reconciling multidimensional policies in the context of overlapping or contending jurisdictions, it is far from clear what the feasible options might be. In Najam's global analysis of policy implementation, he concluded that willingness, concern, and capacity on the part of government are of limited use in understanding what happens at the local level, at which point "the state has already disintegrated into myriad organizations, agencies and actors pursuing different, often conflicting, interests and strategies" (Najam 1995, p. 1).

In many ways threatening, the increasing diversity of voices speaking on a growing range of social issues, with an apparently diminishing sense of common purpose, is probably a function of an also increasingly unpredictable operating environment. Social and economic behaviour and changes in the physical environment no doubt have patterns, but as chaos theory suggests, these are only

occasionally visible and almost never controllable. Policy formation and implementation are thus coming to involve a more diffuse and variable mix of
communities and actors, both within and outside the formal structures of government. These are people active in what Rifkin (1996, pp. 86–87) calls "adap-
tive change," the iterative and interactive, not always purposive, reformulation
of policy in terms of its expected and potential impacts on their own interests
and needs. At a macrolevel, the communities affected by policies form the wide
"networks of interest" engaged in discussion or debate on a given issue. On a
mesolevel, they are the more specifically focused coalitions and public, private,
and civil organizations that, having reasonable consensus within themselves, can
hold often sharply contrasting views between them (Lindquist 1992; Lee 1993).
At the microlevel, and often ignored in the policy-making paradigm, they are
the social institutions (community groups, families, nongovernmental organizations) that try to cope with policies made elsewhere but, with greater frequency,
are coming together to create their own alternatives.

The concept of integrated-policy formulation and implementation has
been evolving since the United Nations Conference on Environment and
Development suggested a new approach to the actors and the complex of problems they face as nations, communities, and individuals. The aim of integrated-
policy development is to better understand the scope of the changes occurring
and to create more democratic mechanisms to frame problems and reach consensus. The process can be seen as a cooperative search for a negotiated path to
more sustainable, equitable, and inclusive governance of social, economic, and
environmental systems. Its emphasis is on realizing shared "values, objectives
and frameworks of interpretation, in the context of which a multiplying range
of players can devise a continuing succession of more particular 'solutions' and
mutually can co-ordinate their actions" (Rosell 1995, p. xi).

Integration as a concept thus defines *inter alia* a process of complex learning, one involving many different individuals within the contexts of their diverse
policy and social "systems," considering and changing how they see and act on
the world. As an approach to policy-making and implementation, therefore,
integration will succeed only to the extent its proponents recognize that,
although it is true that individuals and organizations may need better ways of
knowing, adapting, and reflecting on their actions, the learning required to
achieve these goals is inherently risky, difficult, and fairly chaotic. Developing
the skills, knowledge, attitudes, and behaviours needed to cross the boundaries
between policy perspectives and positions requires considerable intellectual and
social space, commitment, and resources. Some of this learning will happen naturally as people cope with changing circumstances, but much of it will need to
be facilitated.

The purpose of this chapter is to begin to explore aspects of learning that
seem especially relevant to the idea of integrated-policy formulation and implementation. Specifically, it is concerned with the goals of addressing multisectoral

problems and increasingly chaotic governance systems in more sustainable and participative ways.

At the appropriate stages, we suggest one or more propositions intended to focus the preceding discussion.

> **Proposition 1**
> Interventions to develop integrated economic, social, and environmental poli-
> cies and the structures and processes to sustain their implementation should be
> understood fundamentally as learning events. These learning events enable
> people from diverse communities to acquire information, develop skills of analy-
> sis, challenge old values, and adopt new attitudes.

Concepts of learning

Most broadly understood, learning is the process by which people make sense of their social and physical environments in progressively more intricate ways, derive meaning, develop guiding principles, and become better able to predict the consequences of their own and others' actions. Learning enables human development. Increasingly accurate and useful knowledge, skills, and values evolve as we encounter data; organize, refine, and contextualize them; and inter-nalize them to create new capacities. Learning results in relatively stable but cumulative change. Previous learning remains and influences one's outlook but is permanently changed by new learning.

Learning is both an individual and a social activity. Genetic inheritance at least partly determines its potential. Its realization, however, is largely a function of experience and is motivated and influenced by one's sociocultural and physi-cal environment, including the actions and perceptions of other people, the distribution of power in and between groups, and the learning event or task itself — its complexity, urgency, and novelty. Also, learning is natural, motivated by the need everyone has to make sense of the world. It happens with or with-out intervention by outside agents. Referred to variously in terms of cognitive structures, mental maps, and explanatory patterns, this process of making sense of the world occurs as we seek to establish relationships among the disparate pieces of our experience. We attempt to make the process of dealing with our environment more efficient by creating "hooks" — explanatory concepts — to interpret and manage it. These are the structures we use to put our experience into a particular perspective, expressed through our attitudes, beliefs, values, philosophy, and religious convictions.

As suggested above, although people seek patterns in their experience and make cognitive maps for themselves, they also do this for and with others. Our ability to live in groups is essentially a function of our capacity to generate and develop common explanations — shared maps. Learning is thus affected by and affects our surrounding social and institutional culture. Understood as myths, cosmologies, or learning metaphors, these public explanations affect not just what is learned but also who participates in learning, how learning occurs, and whose learning counts (see Thomas 1989; Michael 1992; Rosell 1995). The contexts that influence what and how we learn may be broad, cross-national cultures, such as those expressed in philosophical or religious concepts. But they may also be the more defined cultures at provincial, state, ethnic group, local community, or organization level. These contexts govern expectations, not always consistently, about how people should think and behave.

This mix of internal and external determinants of learning is important in the context of efforts to change policy. People working in integrated-policy development face differing degrees of difficulty and are inclined to different solutions, as each participating group's unique personal, cultural, and social histories influence that group's attempts to accommodate experience. This process of accommodation is, in effect, learning. How that learning happens at the level of both the individual and the group determines whether participants develop the abilities to collaborate, test boundaries, alter designs, identify and weigh options, assess comparative advantage, and synthesize their experience.

Two concepts are particularly useful in understanding the role of learning in facilitating change among both individuals and groups. The first concerns the three different "loops" of learning people use to detect and correct error in their environments (Argyris and Schon 1978). Through simple, single-loop learning, we learn from experience to adjust our behaviour, for example, by rewording a policy that has proven ineffective. In more complex, double-loop learning, negative feedback from experience leads not just to adjusted behaviour (or changing written policy) but also to one's examining the basic assumptions of behaviour, reconsidering one's underlying rationales and assumptions. Finally, through deutero learning, we learn how to learn and begin to understand the process of learning and the behaviours and strategies that inhibit or facilitate that process.

The second element of Argyris and Schon's (1978) framework is the concept of people's "theories of action." Derived from conceptual categories and cognitive maps, such theories are essentially the logic individuals use in devising, applying, and reforming the working rules to govern their own behaviour and to enable them to interpret the behaviour of others. We use these theories to decide which knowledge, behaviours, or attitudes are important and acceptable and which are not.

The two concepts are related. We learn our theories of action while we construct and revise our mental constructs about how things work; in turn, they influence subsequent learning by determining which things we pay attention to and on what basis we are prepared to act. Theories of action appear in two forms, which do not always match: "espoused theory," or the rules we **define** as governing our behaviour and expectations; and "theory-in-use," or the rules we

actually apply. An espoused theory may be, for example, that we should and will learn from criticism or from new information; our theory-in-use, however, may be that criticism and negative feedback are threats to be countered by defensive argument, disengagement, and a shutdown of the learning process. A theory-in-use that makes the exposure of problems or of underlying assumptions culturally unacceptable is less likely to produce the kind of meta-analysis (deutero learning) needed to dismantle the often high walls that sector-based policy-makers build to shut out significantly different ways of sharing power.

A final general point about learning is the somewhat self-evident fact that most of it happens in unplanned or incidental ways as people negotiate their way through the world. Unplanned learning occurs in any context, including planned interventions, and can either complement or contradict formal learning objectives in a given situation. Thus, for example, some participants in the policy process may have learned to be wary of or defensive about information or directives they fail to understand or that appear to contradict their basic values or core beliefs. They may apply that learning by remaining passive. Others may have learned to avoid taking risks beyond their immediate capacity to manage them and so may disengage from the policy process or make extraordinary resource demands on it. Some may simply misinterpret the implications of an innovation, failing to see it as different, but instead seeing it as reconfirming old patterns of belief and behaviour.

From the perspective of promoting change, such learning is neither good nor bad. It is functional or dysfunctional to the extent that it enables participants in the process (the community group, the round table, the organization) to move forward. Learning is a risky and sometimes threatening road to change, and people are less likely to begin the journey without some sense of direction and of belonging in the process. Thus, interventions are more likely to work if the disjunction between new ideas and existing ones is great enough to draw attention to these new ideas but not so great that people cannot imagine implementing them. Interventions are also more likely to be positive if they are user centred, taking into account individual differences in culture, experience, and status.

In sum, policy innovations will succeed or fail, be renegotiated or redirected, for reasons less to do with their merits than to do with how effectively they engage people in the learning needed to notice, assess, interpret, and apply these innovations — in other words, to make them their own.

But learning is not only about change; it is also about maintaining a balance between reinforcement of old learning and creation of new. On the one hand, learning allows us to resist external challenges and maintain a sense of stability in our experience by enabling us to assimilate new experience into existing frames of reference when the match is good between the two. On the other hand, learning permits us to adapt those frameworks to accommodate and reassemble experience when the match between experience and old frameworks is poor or unhelpful. The act of integrating a number of different policy perspectives implies this continuing alternation between "equilibrium and dis-equilibrium, between stability and change," which results in "high levels of understanding" (Alvarez 1994, pp. 153–154). This alternation is a necessary feature of all learning, whether spontaneous or facilitated; it enables us to achieve both the radical paradigm shifts and the sustained application of new patterns of thought needed to develop negotiated policy decisions.

Proposition 2
A variety of cognitive structures are involved in the learning, values, expectations, and behaviours of the groups involved in an intervention. Interventions must take these differences into account to effectively anticipate the stages of learning that will be needed to achieve and sustain consensus among people whose interests and priorities diverge.

Proposition 3
To achieve and sustain consensus on complex integrated-policy issues, the experience of those participating in the policy process must be congruent with stated goals, because the most powerful learning is derived from experience.

Facilitating change through learning

Social cohesion involves building shared values and communities of interpretation, reducing disparities ... enabling people to have a sense that they are engaged in a common enterprise, facing shared challenges ... . In [a] rapidly changing environment ... social cohesion needs continually to be constructed. It depends fundamentally on the capacity of individuals and social institutions continually to learn and to adapt together.

Rosell (1995, p. 78)

Multistakeholder negotiations bring together groups and individuals with often significantly divergent experience, priorities, and access to resources (including power) and major differences in their understanding of issues and priorities. The nature of change implied by these interventions cannot be mandated or controlled. The learning involved in this situation is influenced by how well interveners account for these differences among participants and by participants' readiness to become engaged in and sustain their engagement in this process.

Readiness

Readiness, as a concept in learning theory, recognizes that the ways in which people engage with their environment depend on what they bring to it: their knowledge and sense of self; the cognitive categories and explanatory maps they have developed from previous learning; their capacity to apply different loops of learning and their theories of action; and their willingness to take risks. At the level of a group or organization, the concept of readiness can also be applied to the ways people share and manage their knowledge constructs and strategies for learning.

A first concern in initiating a process of change is whether people are ready to begin the process: the motivation they have to turn up and to stay involved. Research suggests, for example, that an "opportunistic" motivation for adopting a policy or innovation — that is, one compelled by authority or induced by the promise of material reward — leads to no meaningful or sustainable implementation of change. Conversely, a "problem-solving" motivation, derived from a genuine appreciation of the necessity for change, is more likely to lead to the kind of engagement needed for serious consideration, adaptation, and integration of policy interventions (Berman and McLaughlin 1976).

This makes sense in terms of learning — a serious and uncertain business if it exposes assumptions and challenges fundamental concepts. Motivation to participate in complex negotiations needs to come from a belief that the issue is important and a belief in oneself as a potentially effective agent in the process. Concepts used in multistakeholder processes of ownership and buy-in explain the importance of such motivation: unless people feel in some degree in control of an initiative, understand its rationale, and accept responsibility for influencing its direction, they are unlikely to expend the effort for the critical reflection and skills development needed to make it happen.

In promoting policy integration where the aim is to generate or strengthen the mutuality of interest that leads to joint action, this motivation also involves a willingness to create and maintain the social learning unit (Finger and Verlaan 1995) — to develop the attitudes and skills of cooperation, communication, and exchange; a common knowledge base and sense of purpose; and the strategies for double-loop or deutero learning that underpin effective social organization. Broadly based democratic structures, such as universal suffrage, provide people with a certain level of influence. But the practice of negotiation is more likely to develop in much smaller communities or coalitions, in which the minorities care enough to organize themselves to defend their own interests in the policy debate and, in the most constructive of policy processes, learn from each other (Lee 1993).

Coming to care is, therefore, a critical part of the process of change.[1] Unfortunately, people often fail to see a commonality of interest that is sufficiently compelling to motivate them to participate much beyond turning up. Different professions, levels of education, cultural and religious backgrounds, even management and working styles, can generate sufficient variation in expectations and perspectives to colour how a social problem or policy decision is

---

[1]Armstrong, G. 1995. Integrating policy: implementing organizational change. International Development Research Centre, Ottawa, ON, Canada. Unpublished manuscript. 44 pp.

perceived, what information is contributed and how it is valued, and the range of alternatives people are willing to consider.

Recognizing differences in readiness and providing means and incentives as needed to motivate people to care are important elements in change intervention. In such inevitably uncertain, nonlinear, and time-consuming processes, methods used to initiate negotiations need to be diverse to respond to, and build from, the specific interests, experiences, and perspectives of the people involved. Because the inevitable differences in cognitive structures and learning styles are rarely immediately visible, beginning and maintaining a process of change are most effective if "space" — in the form of time, legitimacy, and facilitation — is explicitly allowed for the expression of these differences. Effectiveness at this stage also depends on methods being open-ended and iterative, rather than predetermined or highly structured; on respecting the participants' need both to confirm and to challenge existing patterns of learning; and on giving participants the opportunity to test the parameters of the venture and decide whether to engage in the process.

Taking into account the literal and figurative language of the intervention, the values it reflects, and the norms constituting what and how learning should take place is also critical in supporting people's willingness and ability to engage in an integrative process of policy development. Multistakeholder negotiations and public commissions tend to be organized in literate, cosmopolitan language, based on the paradigms and strategies of the most powerful social actors, and presented in very public ways. Indigenous and foreign cultures, the more marginal subcultures of dominant sociopolitical systems (often the groups who matter most in the effort to better understand and change approaches to basic social problems), are thus often the least able to participate effectively.

Readiness is also affected by others who are involved. Senior managers, political leaders, and public-opinion setters — those who have a direct interest in the maintenance, exploitation, or change of governance processes — must be, and be seen to be, actively committed to the policy-change process. Transparency and common purpose reduce the risks of learning. In some cases, this means making it clear that there is no other choice — for example, a clear message from government that "change will occur with or without one's participation" (Owen, this volume). But the active participation of leadership groups is needed because it implies the regulatory and resource follow-up, transparent agenda, and serious longevity of the exercise that are necessary conditions for convincing and broad-based commitment.

Governments and senior managers also constitute interest groups, of course, with their own stages of readiness for learning and change. Unfortunately, practice suggests that they are the most difficult interest groups to engage in the learning process, reflecting perhaps a leader's heightened sensitivity to the risks that consultation poses to power or to priorities in other areas of public policy. Commitment to learning among leadership groups is nevertheless necessary if policy-making processes are to change in durable ways.

A final implication of the idea of readiness is that those expected to be active in a change or learning process will be significantly more effective in doing that if they clearly recognize themselves as engaged in the task. Unplanned, unconscious learning always happens, but it is not always the most efficient or effective way to realize specific goals, either our own or those set for us. This is especially important in the context of creating new political and civil arrangements for integrated policy. These efforts are far more likely to result in shared, higher-order learning-to-learn capacities if people set out purposively in the work of designing, evaluating and implementing new arrangements. Drawing on the experience of participants in an international training program aimed at strengthening integrative policy-making, evaluators concluded that this type of learning

> *cannot take place unless the learning unit [organization, community] itself consciously becomes a variable. It is only possible to learn one's way out of a vicious circle [of failed governance] by identifying and challenging the constraints of that circle. Learning per se becomes the organizing principle for the unit ... [where people are] willing to do constructive damage to the status quo.*
> Finger and Verlaan (1995, p. 511)

As suggested by the Commission on Resources and Environment (CORE) case (see Chapter 4) and the Ontario Commission on Planning and Development Reform case (see Chapter 5), it seems that if no conscious sense of sharing or at least a modicum of common cause exists, people are unlikely to make the kind of paradigm shift required to move from simple social consultation, through bargaining trade-offs, to comprehensive and sustained social change.

Proposition 4
People learn different skills at differing speeds, and their new learning will be informed by past experience. Consequently, they will differ in their readiness for learning. Interventions should attempt to assess these differences, use various means of motivating interest, and accommodate variations in the pace and timing of their design.

Functions of conflict

As suggested earlier, a critical motivation to engage in learning arises from confronting a negative: a social or environmental condition that is no longer tenable; or information that contradicts existing assumptions or reveals an incongruence between one's espoused theory and his or her theory-in-use. Conflict in a multisector group can motivate when it serves as a brake to the status quo by exposing the actual heterogeneity within the society and thus reveals the "extent to which knowledge is contested" (S. Klees in Rosell 1995, p. 87). Most of the experience reflected in this and earlier research indicates that the initial motivation to engage in integrative processes, at any level, is a function of growing frustration with current practice — if not actual conflict — or of the need to respond to new requirements for action, such as environmental-assessment legislation.

In this sense, conflict can be an important and constructive part of the social-change process. Most significant learning occurs with a challenge to existing attitudes, values, and beliefs, giving individuals and groups the opportunity for reexamination and reformulation. Indeed, for double-loop or deutero learning to occur, basic assumptions must be cast into doubt, and errors must be detected. But maintaining stability is also a natural and necessary dimension of individual and social learning. People and institutions never change easily, and this is for a good reason: the "reluctance to tamper with [work systems] ... should be construed not as narrow mindedness, but rather as a preference for maintaining a successful formula" (DePape, this volume, p. 77).

Confrontation and contradiction are, therefore, delicate matters: they are necessary to learning, but they can also shut down the learning process if they overwhelm people's capacities to listen, negotiate, and adapt. Advocacy or policy coalitions and groups organized for policy review or community action may share core beliefs only to a degree. Cognitive structures and mental maps derived from different learning experiences may be only more or less complementary and therefore only more or less able to govern a learning-for-change exercise consistently or coherently. Policy coalitions may express common commitment to the quest for solutions to specific problems in the public arena, but it cannot be expected that these coalitions will embrace each other's basic belief systems. The more successful the intervention is at exposing divergent core values, the more it may cause serious conflicts to emerge.

Learning is itself occasioned by situations of conflict at some level; however, the capacity to learn provides the necessary (though not always sufficient) means of resolving those conflicts through processes of disaggregation (analysis), hypothesizing (scenario-building), reconciliation (synthesis), and reintegration (creation of new constructs and patterns). Multiple and contending values will facilitate reflection and inquiry where there is a conscious understaking to use the opportunity to compose problems in different ways and actively seek new explanations (Wright and Morley 1989).

The need for a new language

People trying to reach a consensus need, first, to search for an agreed language (concepts, frames of reference, points of departure) to bridge differences in beliefs and facilitate dialogue. When coalitions clash, it is important to determine some level of common values, however peripheral these may be to their core beliefs. As described earlier, such shared values are explanatory constructs that individuals can use as hooks to make connections with their own cognitive categories and to help create a common mental map. On this basis, participants in a dialogue can establish the credibility of the data, that is, the theories-in-action, or rules of the group, which are critical in moving discussion to higher levels and in identifying new options.

The concept of negotiations as "conversations" among communities or interest groups affected by a policy is important to understanding and managing multistakeholder mechanisms. The ways in which groups with different experiences and different levels of power converse will determine, to a large extent, whether conflict is functional or dysfunctional in building an understanding. Change results from exposure, over time, to a pluralistic decision-making environment, with conflict as one dimension, one response, to that conversation.

Learning ultimately reflects political reality

It is critically important that the espoused theories of people who intervene to foster integrated processes be congruent with their theories-in-use. Interventions proclaiming the intention to create multistakeholder consensus on policy, not simply to promote the old regimes dressed in a new rhetoric, actually need to follow through with this. If the espoused theory recognizes the value of diversity, of opening the negotiations to dissent and to new or unexpected policy developments, but the theory-in-use maintains a central, top-down, and highly scheduled control of the process, the latter can be expected to prevail.

In these circumstances, people are less likely to learn to accommodate the stated new goals of the intervention and more likely to reinforce existing strategies for defending themselves against external controls. They will likely learn to mistrust and be closed to other perspectives; to expect answers from experts, rather than from themselves; and to think of themselves as being, in fact, no more empowered than they were in the past, as business proceeds as usual. Such incidental learning may ultimately prove dysfunctional in the long term, undermining rather than sustaining the change process, exacerbating tensions by putting people through a frustrating public exercise, and casting doubt on the credibility of the entire negotiation exercise.

For integration to work, the process of multiparty negotiation must not only appear to be genuinely participative but actually be so. It needs to generate trust and be transparent. It must reinforce two ideas: that the participatory process is not simply a smoke screen to hide someone's intention to proceed with another agenda and that all participants are genuinely committed to finding common ground. The process must provide opportunities, over a sustained period and in iterative ways, for dealing with conflict and contradiction by publicly seeking better information, formulating and testing hypotheses, examining assumptions, and rethinking values, attitudes, and behaviours. These are skills central to broadening the community of people able to intervene in the questions of governance, helping them to understand different perspectives and policy options and to see the world in more comprehensive, less linear, conflictual, or win–lose ways.

Proposition 5
Incidental or unplanned learning will occur during any intervention to achieve and sustain an integrated policy and may be crucial to its success. The impacts of such learning can be made more positive by ensuring congruence between the goals of the intervention and its methods.

Proposition 6
Heterogeneity can be a positive force for learning during multistakeholder negotiations, but it can also foster conflict. Conflict, however, can serve as a useful focus for negotiations in developing and sustaining complex integrated policies if it is governed by norms of mutual respect and forms the explicit content of the issues being negotiated.

A margin for learning

Individuals and groups have options when confronted with pressures to change. They can experiment, allowing themselves to challenge closely held beliefs central to their world view, or they can retreat, perceiving the occasion not as an opportunity but as a threat.

Studies on the adoption of innovations, together with learning theory, indicate that people are more likely to engage in change, challenge assumptions, and manage innovations if they have, and see themselves as having, the space to take a risk: the intellectual, emotional, social, and basic life supports needed to sustain themselves through the experiment, especially if it fails. Learning requires room to fail, without actually endangering one's life, welfare, or intellectual and emotional stability. Referred to sometimes as "margin," it is a person's sense of the space he or she has for suspending established patterns and learning something new.[2]

Margin is a necessary condition for questioning, testing, and revising ideas and skills. Without it, people are unlikely to have the motivation, intellectual resources, or tolerance for ambiguity needed to change direction. Margin recognizes that any innovation, whether a new policy, technology, or idea, makes demands on those expected to accept and implement it. Innovation is disruptive (Zandstra et al. 1979, p. 254):

> *The equilibrium attained before the introduction of the new activity is disturbed. Because new ideas are generally based on ideas and experiences generated from outside ... an adjustment and adaptation period is usually required to determine what changes are required ... to examine these changes in the context of the resources available to the [receiving community].*

---

[2] Bernard, A.K. 1990. Issues in research: learning, education and ethnography. International Development Research Centre, Ottawa, ON, Canada. Unpublished manuscript. 92 pp.

In most cases of life change, individuals or groups already have adequate margin for learning. A major problem for many interventions aimed at producing changes to solve seemingly intractable and complex social problems is that the individuals, communities, and institutions central to the process are often precisely those that have little or no margin for experimentation. They may already be at the limits of physical, financial, or intellectual resources; in extreme cases, literally on the borderline for survival. Reactions from these groups to the idea of substantially rethinking their paradigms and behaviour are more likely to be ones of reservation and rejection than engagement. They simply judge the risks to be too high.

Providing margin for risk-taking in multistakeholder negotiations tends to level the playing field and enable genuine partnership. Providing resources to the less powerful groups may permit them the time and flexibility to experiment with new processes and to develop confidence and skills in managing them; it can also forestall the superficial participation or premature closure that tends to come with unequal relationships (Owen, this volume). Interventions can create the margin available for learning and experimentation by providing, for example,

+ Social or political legitimacy for the process at the highest policy or management levels;

+ Information needed to demystify the process, clarify the risks, and make the uncertainty less threatening (Owen, this volume);

+ Facilitators who have both the knowledge of the context and the innovation and the skills to help people make constructive links between them;

+ Time to ensure that all views are heard, that changes in content and scope of positions are fed back into the process, and that anxieties about competing agendas are mitigated (for example, through child care or work-release time);

+ User-specific tools and training to enable affected people at all levels to disaggregate broadly stated problems, to collect and categorize pertinent data, to analyze, interpret, and evaluate implications, to determine ranges of appropriate action, and to pass on their learning experience to colleagues down the line (DePape, this volume); and

+ Where appropriate, the practical resources of income and livelihood protection needed by those who are asked to change their jobs or reconfigure their community structures.

Margin can also be provided in the design of the intervention, with a balance between having sufficient structure to encourage people to explore new ideas and giving people sufficient freedom for spontaneous capacities to emerge. The mechanisms of change are important in creating "a neutral and common or middle ground in which the various interests ... [can] establish a collaborative approach" (Morley 1989, p. 182).

Examples of this are the various special units, coordination committees, and working groups or committees designed to be spaces for change. Such arrangements have been described as "buffer institutions," formal and informal agencies "designed to facilitate the transitional phase" between traditional and modern technology and between indigenous and Western knowledge. They aim at "softening the constraints" and ensuring the flexibility needed for effective adaptations (Zandstra et al. 1979, p. 254).

The design of an intervention should also provide a sufficiently clear framework or schedule for the activity, along with a public agenda with directions for realistic and acceptable solutions (DePape, this volume; Owen, this volume; Penfold, this volume), and balance these with sufficient training and information to push that framework to its most inclusive limits and, presumably, even to question its validity. The design of the intervention needs to blend top-down principles with bottom-up public negotiations (Fullan 1996; Owen, this volume).

Finally, it would be a mistake to assume that only the poor or disenfranchised require margin for learning. Those with power have a similar need to perceive room to manoeuvre; resources to mitigate risk when experimenting with policy innovations; and ways to ensure that changes will not diminish their ability to manage their position and priorities. Margin for learning is generally greatest at the periphery of people's belief systems, where core values are not threatened and basic cognitive structures are not under immediate attack (Lindquist 1992). However, some people may recognize the margin they have or feel ready to use or enhance it only if their basic interests are threatened. Sometimes, only when the fundamentals of assumptions and behaviour are challenged do people in positions of power see that there might be previously unrecognized room to manoeuvre and to mitigate the consequences of change.

This may help to explain the motivation of political leaders who agree to experiment with radical policy innovations only after they have perceived that threats to their power base are greater from leaving problems unresolved than from opening a consultative process. Such perceptions can be usefully and constructively encouraged through well-facilitated interventions from other agents in the broader community, including the use of a public challenge. Confrontation can lead to the conclusion that less risk is involved in innovation than in trying to maintain the status quo. Armstrong,[3] in this regard, reports the experience of two out-of-government practitioners and their perspective that "governments have to realize that they risk being left behind if they do not participate"; "when the state is not able to react properly, it is more willing to accept alternatives."

_____

[3] See p. 20 in Armstrong, G. 1995. Integrating policy: a matter of learning. International Development Research Centre, Ottawa, ON, Canada. Unpublished manuscript. 46 pp.

Proposition 7
It is important to recognize people's margin for learning and experimentation and to differentiate between their core (and less flexible) interests and beliefs and their more negotiable, secondary interests. Policy innovations and new implementation strategies will be more readily accepted if interventions focus on areas of common accord in people's belief systems or in policy coalitions' core values.

Incremental learning during negotiations

It is important to recognize that interventions provide only the space and occasion for people and groups to undertake new learning in new ways; interventions are not ends in themselves.

Although margin is a condition for learning throughout the process of change, the need for interventions to create it should diminish as participants become better able to create their own through improved double-loop and deutero learning, communication patterns, and support networks. Taking the policy-innovation process in incremental steps facilitates people's ability to create their own margin, reducing the degree of risk and conflict by making each aspect of the proposed change more open to analysis, influence, and manipulation. Staggering an intervention's progress can encourage successive realizations of the different dimensions of broadly stated goals, gradually clarifying and contextualizing them. This mitigates the potential failure of an all-or-nothing, forced consensus (Owen, this volume) and allows an increasingly wider spectrum of engaged participants to adjust to, and sometimes adjust, the innovation.

An incremental approach recognizes the complexity and uncertainty of integrated-policy innovations and the fact that "we still lack a good understanding of genuinely well-integrated policies" (Penfold, this volume, p. 170). Not only is the correct route to an ideal not necessarily self-evident, but also "strategies for generating information are not clear" (Penfold, this volume, p. 172). They are certainly not linear. Working toward consensus becomes, in this context, more a matter of searching for meaning than of producing an analytically generated solution (Penfold, this volume). Flexibility and iteration will allow people to gradually map the problem as issues emerge from the process (Rifkin 1996).

The "search conference," as an approach to cross-sectoral policy-making, evolved from just such an idea of progressive synthesis. In this exercise, multiple strands of information and perspectives are brought together to create new "explanatory designs" (Wright and Morley 1989, p. 220). Search conferencing defines a progressive, "spiral" approach, based on three continually evolving processes of learning: imaging — "using our patterns of experience to look for meaning and 'likeness' in the situation"; presenting — externalizing and communicating these images so they can be tested; and testing — to "provide feedback on how we are doing and 'feedforward' on how to refine our image."

Innovations that can be divided into different components are more readily amenable to this type of interactive accommodation and thus more easily accepted than those that are immutable, win–lose, or "sturdy." Dealing with identifiable (albeit interdependent) dimensions of a complex problem allows people to negotiate between new ideas and old and to gradually adapt their cognitive structures and operational paradigms. Breaking the learning process down into incremental steps helps build consensus, strengthen communication between groups, and increase both understanding and trust in the feasibility of multiparty negotiations. Like learners in any situation, parties to such negotiations require positive reinforcement — the small triumphs of concrete and substantive achievements to confirm their sense of their ownership of the process and to give them something to take back to their constituencies.

An incremental, iterative, and conversational approach to negotiations allows participants to test ideas in the context of their regular lives and so more transparently and congruently consult and participate with their constituencies and colleagues. Breaking negotiations into workable increments gives people time to learn about their own and their opponents' core values and interests and to learn where in this spectrum accommodation is possible.

One critical implication of an iterative, incremental approach is that the process becomes at least as important as, if not more important than, the product. The quality and extent of participation at all levels are as central to success as the final document. This implies that the most influential groups, including government and other organizational leaders, need to be involved at the outset and stay the course. When influential groups decide to remain on the periphery of the public dialogue and to commit none of their own resources or power in the search for alternatives, they risk missing important but subtle shifts in the conversation and therefore failing to internalize the logic of arguments developed during the process. Presented with a *fait accompli* at the end of a negotiating process in which they have not participated, leaders are often fearful of moving the experiment further and are thus unlikely to adopt or effectively implement the negotiated policy.

Proposition 8
Consensus during discussions on integrated policy is more likely and sustainable if the intervention is developed in incremental, iterative stages.

An irony in trying to realize negotiated agreements on policy is that, despite all the considerations and experience so far discussed, people rarely believe they have the time to proceed in a carefully staged, incremental, iterative fashion. Confrontation and action typically and almost by definition take place under the pressure to show results and handle the crisis. Although nondirective, responsive facilitation is critical to successful learning, interveners often feel compelled to jump into a fully formed intervention, unwilling or unable to accept the fact that discussions meander and that change unfolds in unstructured and apparently illogical ways.

Managers and policy leaders, as the advocates and often the sponsors of integrated policy, need to become effective facilitators of learning. However, they will become so only when they consciously recognize the implications of that role and understand what they are asking others to do. As suggested earlier, policy leaders need to agree to engage in learning with the communities (colleagues, staff, clients, constituents) they are working with. They must expect to be challenged, as part of the process; to suffer the uncertainty and risk of shared control; and to adjust their visions to include input from increasingly capable target communities. They need to blend "logic and artistry," develop a "middle-level theory of leadership," and try less to direct change than to capitalize on and foster the capacity of people throughout the system to act as change agents (Fullan 1996, pp. 709–710).

Finding common ground, rather than a complete consensus, is likely to be the most achievable outcome of a negotiation process. Interveners must be prepared to proceed even when unanimity is impossible, making it clear that results can go forward based on general approximations of agreement and that parties should therefore continue to talk through their differences. An ability to cope creatively with unpredictability is especially important as the activity moves out of the first catalytic stages into real-life implementation, when policymakers and the communities affected by their decisions often feel particular pressure to control what is largely an uncontrollable process.

Sustaining change during implementation

This section draws on two primary streams of implementation research: that concerned with innovations in education; and that concerned with complex public policy. Both streams have many points of convergence (Pressman and Wildavsky 1973; Berman and McLaughlin 1976; Fullan 1993).

It is clear from the research that implementation is affected by a range of predictable variables: policy content, institutional and organizational contexts, variable commitments from those charged to act, individual and group capacities to faithfully follow articulated goals, and the resource needs of a myriad of identified and yet-to-be-found clients and coalitions. All of these complicate immensely the effort to follow through. So, too, does the fact that policies will be reformulated as they are variously interpreted and applied. Even agreed innovations are redesigned through continuous cycles of adoption and interpretation — cycles only more or less purposive, coherent, or coordinated — and depend for success largely on the implementors' capacities for double-loop learning and congruence in their theories of action.

The unpredictable and variable nature of implementation is not irrational; rather, it reflects the fact that ultimately a new technology, behaviour, or policy must accommodate realities in the field. Implementation research shows that the ability of an innovation to be sustained is a function of mutual adaptation. Implementation is, therefore, also a process of change and so a learning process. It is most likely to happen effectively if policy proponents try to establish the compatibility of the innovation with the intellectual and cultural characteristics of all the actors involved, support transparent and negotiated conflict resolution, and provide room for accommodation. Without this, an innovation is likely to be co-opted (learned at the level of rhetoric, rather than at the level of substance), lost in established practice (interpreted as confirming existing structures), or just ignored. Recalling the idea of unplanned or incidental learning, one can understand that policy innovation that contradicts local contexts, fails to build on available capacities and resources, or presents itself as a threat can in fact promote dysfunctional learning — responses that directly undermine efforts to bring about change, however well intended.

The same learning skills required in negotiating integrated policies are also needed for their implementation. Flexibility in analyzing and testing alternatives is critical as a counterbalance to the risks people face of losing conceptual structures and predictive analyses that enabled effective action in the past. Because the learning process must be incremental, especially within the highly uncertain context of changing policies, a central question is how to systematize it sufficiently to enable people to manage and continue to buy into it. People involved in multidimensional change are being asked to do a great deal: to analyze, synthesize, and evaluate experience in new ways; to consider alternative points of view and core values; to seek, exchange, and interpret new and often unsettling information needed to redefine and solve problems; and to make decisions in different ways, in riskier environments, within untested parameters. They will participate (Bernard 1991, p. 37)

> to the extent that they choose, cognitively, effectively and physically, to engage in establishing, implementing and evaluating both the overall direction of a programme and its operational details. Choice, in this context, implies not merely an agreement to follow but an active decision to assume responsibility in considering the rationale, implications and potential outcomes of the programme ... [including terminating participation when it is judged dysfunctional].

Not easy tasks — they can only be learned through practice. And practice is self-sustaining, forging new social arrangements of people ready to engage in the debate and build new rules for discussion and interaction, sometimes without looking for any direction from traditional hierarchies. Although such arrangements are also useful in implementing innovation, of course, they can pose significant risks for managers trying to control the process, implying yet further rounds of initiating and facilitating change.

Proposition 9
Interveners must take the time to provide both the learning opportunities and the supports for learning to all members of the policy community, even after achieving consensus. Community participation, a willingness to further adapt the policy, and continued support for learning are essential to effective implementation.

Proposition 10
Unpredictability is an unavoidable condition of complex policy environments. Rather than treating it as a negative factor, however, interventions should provide opportunities for people to learn how to create mechanisms and organizational cultures that can deal with unpredictability. Unexpected learning outcomes should be fed back into the process.

Sustaining change through institutions

Interventions that support only the early stages in the development and implementation of complex policies are likely to fail to establish the sense of ownership needed for durable change. To sustain change after initial implementation, one must therefore ensure that the actors continue to participate in testing and adapting policy decisions and in managing the changes that the new policies generate. Sustainable policy integration, in this sense, is concerned not with fidelity to articulated policy per se but with the commitment and capacity of the policy system to manage change in increasingly satisfying ways, relying on its various communities and institutions to continue to monitor, assess, and adapt.

The concept of institution in this context is an especially important one, defined by North (1994, p. 360) as any form of constraint that human beings devise to shape human interaction. Institutions are the durable informal and formal arrangements that prohibit and permit social behaviours; they comprise the rules we establish as a society to manage uncertainty. Institutions thereby serve as the "framework within which human interaction takes place." They are a consequence of social learning (North 1994) and the locus for future learning, whether they are manifested as organizations or merely expressed as beliefs, customs, and norms.

Institutions are important, therefore, to the way individuals and communities respond to complex, variable policy environments, because institutions provide the context for their interaction. As such, institutions are potentially important to sustaining policy innovation, acting as venues for coordinating multiple points of view. As efforts to catalyze innovations proceed, attention needs to be given to creating or strengthening social institutions so that they are capable of legitimizing and maintaining the dynamics of the process at both central and local levels.

New organizations have been tested in the past decade in a number of political jurisdictions to accommodate the variable and problematic policy environments in which economic, social, and environmental interests conflict. Innovations such as the recently disbanded CORE in British Columbia (CORE 1995), various round-table and search-conference mechanisms throughout the world, and a growing number of community-based intermediary organizations to link government and civil society are experiments in institutionalizing multi-party negotiations and developing more integrative, inclusive policies. Although such innovations are not intended to supplant existing political structures, they can provide relatively stable forums for people to establish alliances and networks as they consolidate interests and negotiate win–win options.

To facilitate durable, integrative policy-making and implementation, these organizations need to be congruent and to develop theories of action consistent with principles of openness and diversity; flexibility in dealing with ambiguity; and decentralization of decision-making, coupled with a central core able to sustain and monitor important organizational themes. Such organizations need to be effective in using strategic vision to detect and correct error, to use it to build on successful local variation and experimentation, rather than as a rigid tool for control (Mintzberg 1989; Fullan 1993).

Evaluation as a tool for learning

The fact that creating and implementing change are learning **processes**, rather than **products** of learning, has implications for how integration is evaluated. The most important outcomes of policy change are largely intangible: broader and deeper capacities for policy analysis and synthesis, increased trust among disparate members of the policy community, greater optimism and higher status for the goals of policy integration, more inclusive bureaucratic mind sets, and stronger community-level management skills.

Evaluation processes and standards that aim for a concrete product and a "magic bullet" to justify costs and prove benefits are therefore likely to be mistaken in how they assess both the progress and the problems of integrated-policy activities. Cost–benefit analyses may be of particularly dubious use if one must somehow deal with the dilemma of

> the costs of not engaging in these activities … . How do you compare the cost of running these [negotiated] procedures and developing policy and writing reports, to the cost of instability in the system, lack of investment when you don't. [4]

Evaluation should make the implicit explicit. For this reason, the design of effective *in situ* evaluation is of the highest importance in interventions aimed at

---

[4] See Owen, p. 29 in Armstrong, G. 1995. Integrating policy: a matter of learning. International Development Research Centre, Ottawa, ON, Canada. Unpublished manuscript. 46 pp.

making policy formulation and implementation more inclusive and open. Where high-profile interventions focus on the final achievement of goals, rather than on the intermediate analyses of the successive approximations and rearticulations of those goals, they overlook both the critical role of learning and the value of using feedback as a fundamental tool in that process.

The important thing is not to try to eliminate mistakes, which are in any case inevitable:

> As a learning exercise, one of your products, your most important products, is mistakes ... and those are not looked on fondly ... so there's a great temptation to pull the plug on investment just at the point when you are analysing the mistakes and what you have learned. [5]

What is important is to develop "enough of a working theory of leadership for change combined with mechanisms for personal and collective reflection, so that you inevitably self-correct" (Fullan 1996, p. 715). Unfortunately, evaluation paradigms appropriate to negotiation processes are conspicuously absent in both the theory and the practice of policy integration. If the potential of change interventions is to be realized, much more work is needed to assess the quality and effectiveness of learning; organizational and intergroup performance; and dimensions of social-change interventions, such as transparency, equity, and empowerment.

Proposition 11
Innovations or policies must be subject to reexamination and reformulation during implementation. The measure of success of a policy, therefore, should not be how closely its implementation matches the original formulation but how closely the reformulated policy meets the needs of those affected and promotes ongoing adaptation.

Conclusions

> Ways must be found of sharing responsibility for managing structural change.... What is required ... is a participative process in which different organizations come together in a joint problem-solving, collective-decision process to redesign the rules of the game and redefine their mutual roles and responsibilities around an agreed definition of public interests .... It should be concerned with designing adaptable systems rather than producing blueprints ... not a single isolated event, but a permanent responsibility for public learning.
>
> Metcalfe (1993, pp. 183, 188)

---

[5] See Thomas, p. 29 in Armstrong, G. 1995. Integrating policy: a matter of learning. International Development Research Centre, Ottawa, ON, Canada. Unpublished manuscript. 46 pp.

Learning

Learning is natural, a continuous process of interacting with and interpreting one's environment. Resulting in recognizable outcomes, such as expressions of new values or knowledge, learning is nevertheless most usefully understood as a process, insofar as it is irreversible and cumulative. New learning builds from existing constructs but renders those explanatory patterns no longer usable in precisely the same way.

Learning involves intellectual and physical activity. It implies intellectual and practical engagement in new ideas, testing them against existing knowledge, values, and attitudes, with the result either of reinforcing old patterns or of generating new ones. It requires margin, a sense of having the time and space to consider novel ideas and experiment with new behaviours. The extent and type of margin needed varies within and between individuals, based on the specifics of their past learning, the constraints and supports of their present situation, and the nature of the learning event. Ultimately, margin will come from within the individual, motivated by the level of crisis or dissatisfaction he or she perceives in an environment and the fact that learning, as an activity, is inherently self-motivating: people seek to make sense of things.

Although learning cannot be coerced or controlled from the outside, it can be catalyzed, encouraged, and facilitated. To succeed in this, interventions must ensure that the learner — the person expected to change — is the determining factor in the design and methods of the change process. Interventions to foster, create, or implement new ways to negotiate policy are more likely to produce substantive change if they take into account the variability, complexity, and ultimate uniqueness of the individual's learning process and his or her readiness to learn.

People tend to learn from what they experience. Thus, if an intervention simply requires views on what a policy should be and how it can be integrated, or if it simply requires that people interact across and with other sectors of the community affected by a policy, its outcomes are likely to be equally simply an exchange of information or a superficial linkage. People will confront and accommodate the substance and deal with the implications of innovations only if they have genuine and sustained opportunities to practice these skills — to test innovations against previous learning, create new designs, and reinforce them through practice. To develop better approaches to policy-making, including new assumptions and paradigms, sustainability needs to be understood as a function of people having the opportunity to learn.

Learning is a social and political phenomenon, as much as a psychological one. Although it happens at the level of the individual, people engage in learning as part of, and for the purpose of remaining effectively within, a particular social, cultural, and institutional setting. In this sense, learning is risky; identity, inclusion, the capacity to act, and access to resources all hinge on a person's ability to understand and be understood in his or her various social environments. As suggested by Armstrong, "the minute you invent an organiza-tion [or attempt to change one], you have automatically stimulated a whole set of learning responses," with individuals and groups attempting to assimilate and accommodate the implications of that change.[6] To understand how society changes, one needs to understand change as learning and learning as a sociocul-tural undertaking.

Participation

Government is a vital but in many ways unequal actor in the policy-change process. Both its capacity to provide the support needed to mobilize, facilitate, and institutionalize new ways of developing negotiations within and among communities and constituents affected by policies — and its capacity to deny, impede, or break the process — are critical. It too, therefore, needs to be engaged in the learning process when it comes (or is brought) to the negotiating table as a critical collaborator in, but not the ultimate arbiter of, the social-exchange process.

Participation implies putting one's power and expertise to the test of pub-lic discussion. Although government can delegate such participation to surro-gates, such as commissions, sooner rather than later it needs to expressly and publicly buy into the process. Government must share the risks and responsibil-ities of common ownership if it is to appreciate the context and the logic of the arguments being made; and it needs to support the inclusion of margin for more vulnerable groups. Government needs also to engage in its own internal learn-ing. This involves assessing the capacities of its agencies to negotiate and inte-grate change, ensuring the quality of congruence between its theories of action as espoused and as applied and seeking ways to create double-loop and deutero learning capacities (see Rosell [1995] for a usefully concrete example of how scenario-building makes such learning possible within government).

Integrative policy-making processes have effects well beyond government, of course. To sustain long-term change, it is necessary to broaden the processes of discussion, negotiation, participation, and education to include marginal groups and youth, those people who must live with the long-term implications of change.

Widening the base of participation also concerns the relationship between the people negotiating change and the groups they represent, the vast majority of whose members are not directly or immediately included. It cannot be only the leaders of policy coalitions, community groups, or bureaucracies who go through a learning process at the negotiation table. So, too, must the members of these groups — those who eventually will be implicated if change occurs. There can often be significant incongruence between the mandate given to interest-group representatives as they enter negotiations and the agreements that

---

[6] See p. 13 in Armstrong, G. 1995. Integrating policy: a matter of learning. International Development Research Centre, Ottawa, ON, Canada. Unpublished manuscript. 46 pp.

emerge — and for good reason. If the process has been effective, the experience of exchange has not simply reinforced initial constructs but changed them.

Realizing a product through negotiation is important, but the application of a policy requires sustaining and extending, in successive and varying operational contexts, the process of creating and adapting the product. Complex sociopolicy change is a struggle of invention, and as each stage develops, representatives need somehow to communicate in practical and participatory ways the how and why of the process to the members of their groups. At each point, it is critical to again establish trust, reconcile conflicts, and build shared paradigms. And this must be done with the same emphasis on readiness, margin, transparency, user-focused tools, and interactive mechanisms and the same concern for synthesis, facilitative and flexible language, and appropriate pacing.

Using such an approach has further implications. The more (and more effectively) space for shared learning is made available across the policy system, the less influential control from the centre will be. But as learning theory indicates, we learn from error, and a system of governance that embraces processes of detecting and correcting policy failure is the most likely to sustain itself in the long term (Michael 1992). Tightly controlled strategic planning is unlikely to be effective in coping with unpredictable systems; even less, in managing them.

Ultimately, policy integration will work only if implementers recognize it as a process of learning, with negotiation of goals, risks, and rewards a permanent feature of this process. Recognizing and involving stakeholders as learners is a necessary condition of governing in complex and chaotic environments. Strengthening capacities for continually recreating the learning-for-change mechanisms underlying integrative innovation is vital. The lesson of learning theory for governance is clearly not that the world is made any more straightforward or controllable by policy intervention but that the individuals, institutions, and systems involved can become increasingly more skilful at adapting to the consequences of intervention, more secure in the expectation of being able to continue the learning–adaptation cycle, and thus better prepared to engage in the kinds of policy negotiation implied by integration.

PART II

NORTHERN PERSPECTIVES

Chapter 3

Environmental Integration at an Electric Utility

D. DePape

◇ ✦ ◇

Introduction

The electrical industry is one of the most important, widespread, and environmentally damaging in the world. Fundamental to modern economic development, electricity is the key energy source for most industries and has a prominent role in the increasingly important electronic and information sectors. Electricity is produced in virtually every region in the world to serve residential and industrial needs. Nearly all electricity is generated using thermal, hydroelectric, or nuclear resources, or some combination of these. The production and transmission of electricity create diverse and substantial environmental impacts. Air emissions from thermal power generation, flooding from hydroelectric power generation, radioactive releases from accidents at nuclear power stations, and clearing of forest lands for transmission lines are sources of environmental controversy and concern around the world. Because of its sizable and often highly publicized environmental impacts, the electrical industry is subject to strong pressure from both government and the public to improve its environmental performance. In response, North American, European, and Japanese electric utilities have introduced extensive and often costly environmental-protection works and programs. These measures have moved the electrical industry to the forefront of environmental integration (integrating environmental protection into the production cycle) and provided models for other industries for incorporating environmental factors into their plans, decisions, and actions.

This chapter examines how Manitoba Hydro, a moderately sized Canadian electric utility, introduced numerous environmental-management initiatives and programs and incorporated environmental considerations into every aspect of its operations over a 6-year period, from 1988 to 1994. During this time, environmental management was transformed from a secondary to a high priority for the corporation and evolved from being the responsibility of a few environmental specialists to being an intrinsic part of day-to-day actions and decisions in all areas of the organization. By 1994, Manitoba Hydro had made sufficient progress in environmental management that it felt comfortable about confirming its approach in writing and informing the public about its efforts in this area. The corporation adopted a sustainable-development policy with 10 principles of environmental management and protection and produced its first publicly distributed sustainable-development report, primarily devoted to its environmental-management performance and programs. The initiatives, policy, and report have been positively received by external groups, including the Manitoba Round Table on Environment and the Economy and the International Institute for Sustainable Development.

Scope of the case study

This case study describes how Manitoba Hydro went about incorporating environmental considerations into it its planning, development, and operational activities and decisions during the 6-year period beginning in 1988. Three aspects of environmental integration are examined:

+ Evolution of environmental integration at Manitoba Hydro, including how it progressed through various functions and layers of the corporation and the types of environmental-management tools that were developed to achieve integration (transmission-line projects, one of the corporation's most geographically widespread and environmentally significant functions, are used to illustrate this process);

+ The factors that provided the impetus and facilitated environmental integration in transmission-line projects; and

+ The types and sources of information used to develop the tools of environmental integration (this is illustrated for three of the corporation's environmental-integration tools, as well as the corporation's sustainable-development policy and principles).

Highlights and recommendations resulting from the case study are presented in the last section of the chapter.

Sources of information

I was a senior environmental policy analyst for Manitoba Hydro from 1988 to 1994, where I worked in the department that spearheaded many of the environmental-management initiatives and programs adopted by the organization. Most of the information and insights presented in this document are based on my experience at the corporation and represent my perspective on what transpired. The analysis and perspectives presented have not been formally reviewed or endorsed by Manitoba Hydro. Supporting sources include

+ Documents produced or used in the course of developing specific environmental-management programs and policies;

+ Informal interviews with Manitoba Hydro staff involved in developing or implementing various integration tools; and

+ Annual reports of Manitoba Hydro and the corporation's first sustainable-development report.

About Manitoba Hydro and its environmental impacts

The environmental-management requirements of an organization depend on its activities and their impact on the environment. This subsection presents an overview of Manitoba Hydro, its activities, and the main effects of its activities on the environment.

Manitoba Hydro is a Crown Corporation owned by the Province of Manitoba. Its mandate is to provide continuous, reliable, and economical electricity to the people of Manitoba. A vertically integrated industrial enterprise, the corporation generates, transmits, distributes, and sells electricity via facilities located throughout the province.

More than 95% of its 5 385 MW of generating capacity comes from 12 hydroelectric plants on four river systems in Manitoba. The remainder is produced by two coal generating stations and 12 small diesel power plants that serve remote northern communities. Electrical power at the generating stations is carried along high-voltage, typically long-distance transmission lines to transformer stations located near major markets. At these stations, the power is downloaded to levels that can be fed into the distribution lines serving more than 380 000 customers throughout Manitoba. Power is also exported to and imported from markets in neighbouring provinces and states.

Flooding and water-regime changes caused by hydroelectric generating projects are the most prominent sources of the environmental impacts associated with Manitoba Hydro's activities. Since the early 1960s roughly 2 500 km^2 of flooding has occurred as a result of hydroelectric projects in Manitoba. Other less prominent yet notable sources of environmental impacts include air emissions from thermal generation; opening up and clearing of largely undeveloped areas for transmission lines; chemical vegetation control along transmission lines and at substations; handling, transportation, and disposal of hazardous materials, including polychlorinated biphenyls; disturbance of stream crossings during construction of projects; and contamination at diesel-plant sites. A summary of the sources of impacts, according to basic and specific activities performed by Manitoba Hydro, is presented in Table 1, along with an indication of the relative severity of the impacts that do or can occur. There are 40 combinations of

Table 1. Sources and relative severity of actual and potential impacts of Manitoba Hydro's activities.

Specific activities affecting the environment	Basic activities				
	Hydro generation	Thermal generation	Transmission lines and stations	Distribution lines and stations	Customer and support services
Water related					
Water-system alteration (flooding, drainage, diversion, dyking, blockage)	H				
Station operation (reservoir filling and drawdown, flow control, river discharges and withdrawals)	M				
Stream crossing	M	L	M	M	L
Discharge of effluent (sewage, cooling water)	L	M			L
Land related					
Locating, clearing sites and rights of way	M	L	M	L	L
Excavation of borrow pits, fill areas	M	L	L	L	
Disposal of debris, ash, and waste	L	M	L	L	
Control of tree growth and weeds			M	L	
Air related					
Emission of particulates, work-site dust		M			
Emission of SO_2, NO_x, water vapour		M			
Noise, electric and magnetic fields		M	L	L	
Other					
Spills or incidents related to transport, storage, handling of hazardous materials and fuels	M	M	L	L	L
Generation of solid, liquid, or energy waste	L	L	L	M	L

Source: InterGroup Consultants Ltd.
Note: H, high; L, low; M, moderate; blank, negligible.

basic and specific activities having actual or potential impacts on the environment. The challenge facing Manitoba Hydro has been to develop and implement environmental-management tools that would make its activities less environmentally damaging without jeopardizing the corporation's ability to provide electricity in a continuous, reliable, and economical manner.

Evolution of environmental integration in transmission-line projects

Environmental integration at Manitoba Hydro has been an evolutionary process. Improved environmental management became a high priority during the middle to late 1980s in a few areas of the corporation and in response to new regulatory requirements introduced by the provincial and federal governments and an all-time high level of public concern about environmental degradation. Over a period of 5 years, environmental integration spread throughout the organization as senior management made environmental stewardship a corporate priority and staff became more aware of the implications of environmentally responsible actions. The integration process manifested itself in growth of environmental-services groups within the corporation; development and implementation of a wide variety of environmental-integration or -management tools to meet the needs of different activities and functions; an increasing tendency for staff to develop new tools on a voluntary basis, without the pressure of regulatory requirements; and the interest of line staff in taking more responsibility for improving their environmental-management practices. The following subsection describes the integration process, beginning with a review of the organizational structure; this is followed by an examination of the integration process, from early planning to operation and maintenance of the transmission-line projects.

Organizational setting for integration

Manitoba Hydro's organizational structure was a key factor shaping the manner and pace of environmental integration.

Companies organize themselves to perform their primary and supporting activities effectively and efficiently. The organizational structures they create reflect the company's mandate, goals, competitive environment, and activity profile. In the case of Manitoba Hydro — a large, hierarchical, functionally defined organization operating throughout Manitoba — the primary activities of generating, transmitting, and distributing electricity and serving customers were distributed among 10 divisions (Table 2). Each division was responsible for one or more stages in the project-development and -implementation cycle, which extends from system and project planning to project engineering and construction and ultimately to customer service and project operation and

Table 2. Organization of Manitoba Hydro's primary activities.

Division(s)	Stages and activities covered (usually by separate departments or sections)
Power Resource Planning	System planning for generation System planning for transmission Project planning and design for generation
Engineering	Project planning and design for transmission lines Project planning and design for converter stations
Construction	Construction planning and day-to-day management for generation projects Construction planning and day-to-day management for transmission lines and converter-station projects
Production North, Production South	Operation and maintenance of hydroelectric and thermal generating stations
Northern Region, Western Region, Eastern Region, Central Region	Project planning, construction planning, and implementation of distribution lines and substations located in their region Operation and maintenance of transmission lines, distribution lines, and substations in their regions Provision of direct services to customers in their regions

Source: InterGroup Consultants Ltd.

maintenance. A further distinction was made for the type and size of project and the geographic area served. The stages of the project cycle, types of activities, and area covered by the 10 divisions were as follows:

+ *Power Resource Planning* — system planning for generation and transmission and project planning for all generating stations;

+ *Engineering* — project planning and design for transmission lines and converter stations;

+ *Construction* — construction planning and building, rehabilitation, and decommissioning of generating stations, transmission lines, and converter stations;

+ *Production and System Operation (three divisions)* — operation and maintenance of hydroelectric and thermal generating stations (one division is responsible for stations in southern Manitoba; one, for northern Manitoba; and one, for coordinating the overall electricity-supply system on a day-to-day basis);

+ *Regions (four divisions)* — project planning, construction, rehabilitation, and decommissioning of distribution lines and substations; operation and maintenance of transmission lines, distribution lines, and substations; and provision of direct services to customers, including an electricity-conservation program (one division is responsible for Winnipeg; one, for western Manitoba; one, for eastern Manitoba; and one, for northern Manitoba).

Although each division is overseen by two vice-presidents, it operates almost as a separate organization with its own management and organizational structure. Within a division, numerous departments, often made up of several layers of supervisors and staff, perform the division's functions, applying appropriate standards, practices, and procedures. Work systems have typically evolved over time to meet the corporate mandate of supplying low-cost, reliable electric power and to fulfil corporate responsibilities in areas such as employee health and safety. In general, because these systems offer a high level of effectiveness in achieving their intended purpose, there is an inherent reluctance to tamper with them. This reluctance should be construed not as narrow mindedness, but rather as a preference for maintaining a successful formula.

This was the context for the introduction of environmental integration. Divisions and departments responsible for activities affecting the environment needed to modify their work systems to include the management of environmental impacts. Methods had to be identified and implemented for incorporating environmental considerations into standards, practices, and procedures. An acceptable balance would have to be struck between traditional requirements and the needs of the new environmental agenda.

To be effective, the process of integration had to flexible and capable of accommodating a diversity of circumstances. Integration tools, practices, and procedures had to be designed to be appropriate to, and compatible with, each organizational unit's activities and work systems. Solutions were needed to enable each unit to continue conducting its primary functions in an effective manner. Methods that adversely affected the performance of basic functions would be unacceptable. The variety of activities and the sizable number of divisions and departments responsible for those activities made environmental integration a substantial undertaking that, of necessity, could only occur in successive steps, over an extended time.

To spearhead this process, Manitoba Hydro strengthened its environmental-services functions. One of a number of services supporting the corporation's line departments, environmental services grew in size and importance from the mid-1980s until the early 1990s. During the case-study period, the corporation had three different divisions with responsibility for developing and implementing various aspects of environmental management:

+ *Environmental Affairs* — To serve as catalyst and contributor to environmental integration for all aspects of corporate activity except those covered by Mitigation and by Workplace Health and Safety (this group reported to its own vice-president for several years, and then to the senior vice-president of Engineering and Environment);

+ *Mitigation* — To resolve grievances related to environmental damage caused by existing generating, water-diversion, and water-regulation projects through mitigation and compensation programs and agreements (after reporting to the Environmental Affairs vice-president for several years, this group later reported to the president); and

+ *Workplace Health and Safety* — To coordinate programs for managing hazardous materials and spills response (this group reported to the vice-president of Personnel and Services).

These groups, which more than tripled their staffs between 1988 and 1991, played a key role in spreading environmental management throughout the organization. They identified the priority areas; helped line divisions to better understand the effects their actions and decisions had on the environment, as well as the regulatory and corporate policy requirements they were expected to satisfy; and provided leadership in the creation of a wide variety of tools for incorporating environmental considerations into actions and decisions.

Achieving environmental integration in transmission-line projects

Manitoba Hydro's experience with the environmental integration of transmission-line projects is the focus of this chapter. After generating stations, transmission projects are the largest, most visible, and most environmentally sensitive projects carried out by Manitoba Hydro. A transmission line carries power from remote generating stations to major urban and other load centres and from one load centre to another. Transmission projects are installed when a new generating station is built, to provide backup power service to users of electricity, and to enhance the power supply to areas experiencing growth in electrical demand, although demand-oriented strategies for managing electric power can occasionally defer or eliminate the need for a transmission project. Manitoba's transmission lines vary in length from less than 50 to more than 800 km, require vegetation-controlled right of way ranging from 30 to 60 m, and pass through a wide variety of natural and human settings, including bogs, streams, rivers, boreal forests, agricultural land, and major cities. The possible environmental impacts of transmission lines include increased access to parks, wilderness areas, and undeveloped areas; loss of aesthetic enjoyment; loss of forest cover; disturbed wildlife habitat and movement; erosion of stream banks and sedimentation of streams; fuel spills; herbicide use; and creation of electric and magnetic fields.

The process and tools for integrating environmental protection into the transmission-line project cycle are introduced in the remainder of this subsection and described in greater detail throughout the rest of this chapter.

Integration process
Table 3 identifies the activities in development and maintenance of transmission lines that could have environmental impacts. The most important of these is the actual siting of the transmission lines, as this determines which environmental resources will be affected by the development of the lines. Wherever possible, the

Table 3. Groups with responsibilities that influence environmental impacts of transmission-line projects.

Specific activities affecting the environment	Key decisions and actions affecting environmental impacts	Responsible divisions	Responsible departments or sections	Relevant environmental support group
Locating and clearing sites and rights of way	Choice and timing of projects Location of project, right-of-way width Clearing methods, timing of construction, refinement of location Actual clearing	Power Resource Planning Engineering Construction	Transmission Planning Transmission Design Transmission Construction	Environmental Affairs Environmental Affairs Environmental Affairs
Stream crossing	Location of project Timing of construction Stream-crossing method Actual construction Timing and location of maintenance travel Actual maintenance travel	Engineering Construction Northern, Western, Eastern, Central regions	Transmission Design Transmission Construction Line Maintenance	Environmental Affairs Environmental Affairs Environmental Affairs
Excavation of borrow pits, fill areas	Choice of sites for fill material Actual construction	Construction	Transmission Construction	Environmental Affairs
Disposal of debris, ash, and waste	Choice of disposal areas, method of disposal Actual disposal	Construction	Transmission Construction	Environmental Affairs
Control of tree growth and weeds	Method and frequency of vegetation control Actual vegetation control	Northern, Western, Eastern, Central regions	Line Maintenance in each region Regional Support Services	Environmental Affairs Occupational Health and Safety (re: herbicides)
Noise, electric and magnetic fields	Location and height of transmission lines Testing program for electric and magnetic fields	Engineering Occupational Health and Safety	Transmission Design Occupational Health	Environmental Affairs Occupational Health and Safety
Spills or incidents related to transport, storage, handling of hazardous materials and fuels	Construction methods Actual construction Line-maintenance methods Actual line maintenance	Construction Northern, Western, Eastern, Central regions	Transmission Construction Line Maintenance in each region	Occupational Health and Safety
Generation of solid, liquid, or energy waste	Choice of line current and conductor material, construction methods Actual construction Line-maintenance methods Actual line maintenance	Engineering Construction Northern, Western, Eastern, Central regions	Transmission Design Transmission Construction Line Maintenance in each region	Occupational Health and Safety (re: hazardous waste; otherwise, not covered by existing groups)

Source: InterGroup Consultants Ltd.

aim is to avoid areas of high environmental sensitivity (such as highly valued and sensitive rivers, wetland and wildlife habitats, and designated wilderness areas) and to take advantage of environmentally desirable opportunities (such as use of existing rights of way or already developed areas). Other key activities affecting the environment are actual line clearing and construction work, especially at stream crossings, and control of tree growth and weeds when the line is in operation.

Decisions and actions determining the effects of a transmission line on the environment are taken throughout the project cycle. The choice and timing of transmission-line projects are established during system planning; the location of the transmission line and the right-of-way width, during project planning and design; the method of clearing, the approach to crossing streams, the timing of construction, and the choice of contractors, during construction planning; the actual clearing and construction activities, during construction; and the methods and frequency of controlling vegetation growth and conducting line inspections, during line operation and maintenance.

Each step in the cycle, together with its associated activities, has distinctive technical characteristics and information requirements. As shown in Table 3, at the time of the study, seven divisions and departments within Manitoba Hydro had a role in some aspect of the transmission-line project cycle. Two were involved in various aspects of system or project planning; one, in construction; and four, in ongoing operation and maintenance at various locations in the province. To achieve environmental integration in all aspects of transmission-line projects, every one of these divisions and departments had to become environmentally attuned. Two different divisions, Environmental Affairs and Occupational Health and Safety, were responsible for supporting appropriate environmental-management initiatives or processes in the transmission-line project cycle (Table 3). Environmental Affairs dealt with the planning functions, and both groups dealt with construction, operation, and maintenance functions.

The process of environmental integration for transmission projects was one of progressively introducing and reinforcing a willingness to consider environmental effects in the divisions and departments involved in the project cycle. This was accomplished incrementally from division to division and activity to activity, over a series of projects. A corporate learning process took place that eventually spread to all parts of the organization. Regulatory requirements were the initial catalysts, prompting sections of the corporation to take action. The divisions and departments that participated early in the process spent much effort to reach a thorough understanding of the reasons for and operational implications of environmental management and to find tools and solutions that enabled environmental integration to develop without compromising basic functions. Senior-management support played an important role in building acceptance of the higher costs that environmental management often entails. As more groups in the corporation gained experience with environmental management, it became easier to gain acceptance for it. Less effort had to be put into convincing and educating staff. Later in the process, committed departments sought to improve the effectiveness of their environmental-management tools.

Environmental integration in transmission-line projects began in earnest in 1988. By mid-1994, all division and departments responsible for transmission-line projects were involved to some degree in environmental management. The greatest progress occurred in those divisions and departments dealing with planning and development of new projects, namely, the Construction and the Engineering divisions. They were the first groups in the transmission-line-project cycle to start dealing with environmental integration. Their activities were directly affected by new environmental regulations, which reflected heightened public concern about the environmental impacts of large projects, such as the transmission lines. Under new provincial regulations, high-voltage transmission-line projects needed an environmental licence before construction could begin. As well, new transmission-line projects extending into environmentally sensitive areas faced the prospect of significant delays, because of public opposition. By mid-1994, thinking about and addressing environmental impacts had become part of the planning, decision-making, and implementation processes for all project design and construction functions. Most of the key staff in these areas realized the importance of incorporating environmental considerations into their activities, understood how their actions and decisions affected the environment, and were familiar with the environmental-integration tools required for their activities. The environmental-assessment and -licencing process was well understood and generally accepted. Environmental considerations were being incorporated into project-site selection, construction planning, and day-to-day construction activities. Some planning and construction staff took the initiative to improve these tools and the effectiveness of their environmental-management practices and procedures.

At the time of the study, the four divisions involved in line maintenance were less advanced. Environment had been factored mainly into the regulated areas of spill response and hazardous-materials management, without extending much beyond this into such areas as erosion protection and vegetation control. In 1993, the Northern Region division voluntarily included environmental considerations in all of its line-maintenance functions.

The division involved in planning the overall transmission system was also in the early stages of environmental integration, exploring how environmental costs and regulatory requirements might be factored into its planning analysis.

Integration tools

Development and implementation of environmental-management tools suited to particular functions and activities provided a focal point for the integration process. These tools are the principal means for factoring environmental considerations into decisions and actions, giving operational meaning to the concept of environmental integration. Integration of a particular activity occurs when one or more suitable environmental-management tools are developed for that activity and used on a continuing basis. An important goal of environmental integration is to have environmental-management tools applied to all of an organization's environment-affecting activities. Although this goal does not take into account some qualitative aspects of integration — particularly the need for fundamental changes in the general attitude of staff toward the environment — it provides a very meaningful and easily measured standard for assessing environmental integration. The essence of environmental integration in an electric utility is to identify appropriate tools for specific activities, shape them to fit the primary activity they are intended to influence, and gain acceptance by the affected department or division for their implementation.

The tools developed by Manitoba Hydro for the environmental integration of transmission-line projects are described in Table 4. The table also shows the project stage to which each tool applies and the sequence of their development among and within project stages. Environmental-impact assessment (EIA) for large transmission projects was the first tool to be implemented. Initially applied to hydroelectric generating stations, which can create major flooding and water-regime impacts, this tool was extended to high-voltage transmission-line projects in 1988, when these projects became subject to environmental licencing. Initially, the EIAs were done after key project parameters, such as siting, had been determined by the planners. The assessments focused on identifying possible environmental impacts of construction activities and on proposing mitigative measures to minimize these impacts. Two years later, in response to external pressure to move environmental integration earlier into the project-planning process, the environmental-assessment process was extended to include project siting, probably the single most important determinant of transmission-line environmental impacts.

After EIAs for large and small projects, seven new environmental-management tools were introduced, covering every facet of the transmission-line project cycle. The Workplace Hazardous Material Information System and spill-response training came into effect in 1989 and 1991 as part of corporation-wide initiatives to comply with regulations in these areas. The remaining tools were introduced voluntarily, without regulatory prompting. Some were used to improve the quality of environmental integration in areas already being addressed; others dealt with new areas. Hazardous-materials training and handbooks strengthened efforts previously initiated to manage hazardous materials. Self-directed environmental assessments extended the application of this tool to lower-voltage transmission-line projects, which did not require licencing. Environmental-protection plans for construction and maintenance of new transmission-line projects provided a bridge between the mitigation measures proposed in environmental-assessment reports and the detailed instructions from field supervisors to those building or maintaining the project. Although limited to northern transmission projects, the northern line-maintenance guidelines nevertheless provided the first set of comprehensive standards and procedures for protecting the environment during maintenance of existing transmission lines. An interdepartmental committee recommended a unique

Table 4. Tools for environmental integration of transmission-line projects.

Environmental-integration tool	Features of tool	Actions and decisions affected by tool	Comments on evolution or scope of application
Integration into construction planning			
EIA for large projects	Comprehensively identifies and evaluates impacts of a proposed transmission line when location has already been already determined Identifies mitigation measures to minimize anticipated impacts, particularly during construction Includes public involvement throughout the process Adheres to regulatory guidelines	Refinement of location Clearing methods, timing of clearing Timing of construction Stream-crossing method Waste-disposal method	First environmental-integration tool developed by Manitoba Hydro, initially applied to the Limestone Generating Station in mid-1980s, extended to major transmission lines in 1988 Initial applications occurred when main project parameters, such as line location, had already been established
Self-directed EIA for small projects	Uses a checklist approach to identify potential impacts and required mitigation measures Adheres to internally developed guidelines	Clearing methods, timing of clearing Timing of construction Stream-crossing method Waste-disposal method	Developed to deal with small transmission project that did not merit or require a comprehensive environmental assessment First used in 1992
Integration into project design and engineering			
Integrated site selection and EIA	Extends EIA for large projects to include site selection Explicitly incorporates environmental and socio-economic factors into site-selection process Broadens range of alternative sites to consider better environmental and socioeconomic factors Includes public involvement throughout the process Adheres to regulatory guidelines	Siting alternatives for project, location of project Clearing methods, timing of clearing Timing of construction Stream-crossing method Waste-disposal method	Recognizes that location of a transmission line has a greater bearing on its environmental impact than any other project feature or activity Can be used to avoid environmentally sensitive sites Required considerable restructuring of the project-design process First used in 1989/90 to develop and evaluate alternative corridors for a proposed transmission line for the Conawapa project
Integration into construction management and line maintenance			
Workplace Hazardous Materials Information System	Computerized database that identifies and describes hazardous materials used by Manitoba Hydro	Use of chemicals	Started in 1991, part of an overall, corporate-wide program

(continued)

Table 4 concluded.

Environmental-integration tool	Features of tool	Actions and decisions affected by tool	Comments on evolution or scope of application
Spill-response training and handbook	One-day workshop on how to deal with fuel or hazardous-material spills in the workplace. Handbook providing instructions for dealing with spills	Method of responding to spills during transmission construction and maintenance	Started in 1991, part of an overall corporation-wide spills-response program and broader spills-management strategy. Intended to include site-specific response plans and initiatives to reduce the prospects of a spill
Environmental-protection plan	Built around a series of air-photo mosaics; describes environmental conditions along the length of a proposed transmission line; prescribes detailed environmental-protection measures at each environmentally sensitive site and for particular types of activities	Site-specific methods of clearing for transmission lines and construction, including stream crossing and waste disposal	Only used on transmission lines. First application was in the construction of a new transmission line. Subsequently extended to maintenance of new lines and decommissioning of existing lines. Copies of plan are provided to construction supervisors and contractors. First used in 1993
Hazardous-materials training and handbooks	Workshop on how to handle, collect, ship, and dispose of hazardous wastes. Handbooks providing instructions for dealing with hazardous materials	Day-to-day construction and maintenance practices	Originally intended only for the Conawapa project, was extended to all projects requiring environmental licences, after Conawapa was canceled. Only the second such program by a electric utility in Canada
Forest-enhancement program	Program of reforestation and other forestry initiatives to replace forest-cover losses associated with corporate activities	Has no direct effect on mainstream construction or maintenance activities but operates in parallel with them	Started in 1993, part of an overall corporation-wide hazardous-materials management program
Northern line-maintenance guidelines	Describes environmental standards and practices to apply during transmission-line inspection and repair and for right-of-way management	Method of line maintenance, line repair, and right-of-way management	Developed and approved in 1993. Followed up by training session for line-maintenance staff, with environmental specialist
Integration into transmission planning			
Environmental costing	Incorporates environmental mitigation, compensation, and external costs into cost estimates used in evaluating planning options	Major projects to develop and their timing	Recommendation was reviewed by senior managers and forwarded for executive approval

Source: InterGroup Consultants Ltd.
Note: EIA, environmental-impact assessment.

forest-enhancement program for offsetting the effects of forest-cover losses from clearing rights of ways for transmission lines. A task force examined ways to valuate environmental impacts and incorporate environmental costs into system and project planning.

Factors influencing development and application of integration tools

A number of factors have influenced the development and implementation of environmental integration at Manitoba Hydro. This section identifies these factors through careful examination of seven transmission-line environmental-integration tools, namely EIA for large projects, self-directed assessment of small projects, integrated site selection, environmental-protection plans, the forest-enhancement program, the northern line-maintenance guidelines, and environmental costing. These tools are highly diverse, cover all stages of the project cycle, involve many line divisions and departments, were developed during various stages of the integration process, and reflect both regulatory-induced and voluntary initiatives.

Table 5 presents the seven tools examined in this section and information on key features relevant to their development and application. The following factors played an important role in motivating or facilitating the development and application of these tools:

+ Regulatory requirements;

+ Top-level commitment to environmental management;

+ Environmental specialists on staff; and

+ Willingness of key line staff to buy into the plan.

Several of these factors are interrelated and contribute to each other. Each is discussed below. This discussion also elaborates on the process of the environmental integration of transmission-line projects.

Regulatory requirements

In the latter half of the 1980s, public concern about the environmental effects of human activities escalated to new heights throughout the industrialized world. Toward the end of that decade, the environment was at the top of the public's agenda for action by governments. A flurry of new environmental legislation and regulations significantly expanded the environmental-management requirements imposed on industry. Court challenges by well-organized and well-supported environmental interest groups were launched to ensure that existing legislation was properly interpreted and enforced. These developments had a profound

Table 5. Factors influencing development of selected environmental-integration tools for transmission-line projects.

Integration tool	Driving force(s) prompting development	Method of development and implementation	How affected line department(s) have been involved	How environmental-affairs specialist(s) have been involved
EIA for large projects	Need to comply with the 1988 Manitoba *Environment Act*, which mandated environmental assessments of larger transmission-line projects	Framework and analysis produced by environmental consultants, in accordance with regulatory requirements Proposed mitigation measures incorporated into project drawings, detailed cost estimates, and instructions for construction supervisors	Little involvement in development of framework Included as members of the internal project environmental-assessment committee that oversees the work of consultants and determines acceptability of recommended environmental-protection measures Provides and reviews project-description information used for the environmental assessment	Promoted need for environmental assessment internally, long before legislation was enacted Behind-the-scenes advocate of environmental legislation Selected and managed activities of environmental consultants and coordinated involvement of relevant internal departments in the assessment Worked to ensure that recommendations of assessment were implemented by appropriate departments
Self-directed EIA for small projects	Corporate commitment to environmental stewardship Desire to have all new projects in the corporation subject to some form of environmental review, not just larger projects for which assessment is mandatory	Framework developed by environmental specialist on staff, with assistance from consultant Analysis produced by line department responsible for project Any needed mitigation measures implemented by department responsible for project	Consulted by environmental specialist during development of framework Produces and follows up on assessment	Coordinates and has lead role in producing framework Encourages line departments to produce assessments and, if necessary, provides technical assistance Reviews assessments produced by line departments
Integrated site selection	Anticipation that environmental guidelines for Conawapa transmission-line project would require integrated site selection and that this would become a norm for such projects	Framework and analysis produced by environmental consultant Environmental and socioeconomic constraints, risks, and opportunities included in site-selection analysis managed by Transmission Design Department	Not involved in development of framework but played significant role in establishing how the framework was used Incorporates the environmental analysis into site-selection work	Promoted need for integrated site selection, long before legislation was enacted Selected and managed activities of environmental consultants and coordinated activities with relevant internal departments involved in the siting process

Environmental-protection plan	Corporate commitment to environmental stewardship Interest in producing tools more specific and useful than environmental assessments for field staff involved in transmission-line construction	Framework and analysis produced by environmental specialist on staff Plans attached to specifications for construction contracts and used by construction supervisors	Consulted extensively by environmental specialists during development of prototype and on subsequent plans Project supervisors and foreperson give instructions to field staff based on measures prescribed for specific locations and situations	Developed idea for environmental-protection plans Prepared the plans, sometimes with assistance of consultants
Forest-enhancement program	Interest in facilitating environmental review of Conawapa project Corporate commitment to environmental stewardship and sustainable development	Recommendations and framework developed by interdepartmental committee Implementation administered by Transmission Line Construction Department, with input from interdepartmental committee and external advisory committee (a first for Manitoba Hydro)	Members of the three committees that formulated the program were given lead responsibility for implementation	Initiated formation of and chaired first committee involved in addressing forest-cover replacement Key member of the second and third committees Let others assume leadership role
Northern-line maintenance guidelines	Corporate commitment to environmental stewardship and sustainable development	Recommendations and framework developed by interdepartmental committee Staff of Northern Transmission Line Section incorporate guidelines into their day-to-day activities	Initiated and chaired committee that developed the guidelines Field staff incorporate guidelines into their daily activities	Key member of committee that developed guidelines Produced guidelines booklets that could be easily accessed by staff Provided training to field staff
Environmental costing	Corporate commitment to environmental stewardship and sustainable development Trends in other jurisdictions requiring utilities to integrate environmental protection into their planning	Recommendations and framework developed by interdepartmental committee Transmission-planning staff expected to develop and consider estimates of environmental costs in their costing analyses	Co-initiated and co-chaired committee that produced the recommendation (forwarded for executive approval) Has lead role in implementing the recommendation	Co-initiated and was key member of committee that produced the recommendation (forwarded for executive approval) Provides technical support for developing environmental costs

Source: InterGroup Consultants Ltd.
Note: EIA, environment-impact assessment.

influence on the corporate community and its adoption of environmental-management practices.

Before this burst of activity, the environment received nominal consideration in most corporations, including Manitoba Hydro, and gained attention only when a high-profile project faced some form of public review or when a highly publicized environmental incident occurred. This changed in the late 1980s and early 1990s for Manitoba Hydro and other corporations. Environmental management and compliance began to appear regularly on the agenda of Manitoba Hydro's executive meetings and became a high priority in divisions and departments whose activities were affected by the new legislation and regulations. It became apparent that one could no longer disregard the environment and that environmental protection would have to become an integral part of doing business, like economics, technical performance, and occupational safety.

As shown in Table 5, new or anticipated environmental legislation was a driving force in the development of EIAs and integrated site selection for large projects. It is then no coincidence that of the seven tools shown, these were the first two to be implemented. Regulatory requirements spurred Manitoba Hydro to take environmental management to a new level in the planning and development of its transmission-line projects and other activities. The *Environment Act* of Manitoba, enacted in 1988, imposed a new requirement: that higher-voltage transmission-line projects would require an environmental licence before construction could begin. A key aspect of the licencing process was the submission of an EIA for the project to Manitoba Environment.

Before the Act was passed, the small group of environmental staff within Manitoba Hydro had been advocating EIAs for all generating-station and major transmission-line projects, noting that this had become a requirement in many other jurisdictions and was in keeping with growing public concern about the environment. Their ideas met with resistance from many line departments, concerned that the assessments might identify new impact-management requirements, adding to the complexity and costs of project development. Between 1980 and 1988, one EIA was carried out at Manitoba Hydro, for a large, high-profile generating station. No transmission-line projects were assessed. The new legislation gave environmental staff the leverage to bring EIAs into transmission-line planning and development, and it required the corporation to conduct EIAs before proceeding with new high-voltage projects.

After the new legislation, EIAs became the norm for all planned major transmission-line projects. At least one or two EIAs were then being conducted for this type of project each year. Assessments were also performed on other projects requiring an environmental licence. Coordinated by corporate environmental staff, most of the EIAs were at first performed by outside consultants with expertise in this area, although after several years in-house staff began to undertake some of the EIAs. Environmental staff and consultants worked closely with staff in the Engineering and Construction divisions, which had the main responsibility for project development, to produce accurate project descriptions and identify feasible mitigation measures. Typically, during the preparation of an EIA, an interdepartmental committee was convened at strategic times to compile input and comments from all the departments with an interest in the project.

After the new legislation was introduced, the affected departments had very little time to buy into the environmental-assessment process for transmission lines, as EIAs had to be done quickly to meet project schedules. The environmental staff and consultants understood the regulatory requirements and proceeded accordingly. At first, staff of the line departments raised numerous questions and concerns about this new process with which they had to comply while still pushing to meet schedules. They believed the environmental studies and associated public-consultation processes were unnecessarily elaborate and time consuming and that many of the proposed mitigation measures were too inconvenient and costly. This created tension between the two groups, due largely to the line departments' lack of understanding of the scope, content, and requirements of the environmental-assessment process. Over the years, this problem was largely overcome as the affected departments became involved in a number of EIAs and their key staff acquired the necessary understanding through practical experience. Some of these people eventually became strong advocates of improved and innovative environmental-management techniques in their own departments and in other parts of their divisions.

The integration of environmental protection into site selection was the single most important measure adopted to address the environmental impacts of transmission projects. Although it was not originally required for EIAs of transmission lines, it eventually became part of the EIA guidelines. Anticipation that it eventually would be a requirement prompted its initial application by the corporation. The first time it was applied at Manitoba Hydro was for the assessment of the controversial 800-km transmission line that was part of the $6 billion Conawapa project. This large-scale, high-profile project was designed to deliver power to the Winnipeg area, in southern Manitoba, from the proposed Conawapa generating station on the Nelson River, in northern Manitoba. To reduce risk of encountering problems in the EIA of this project, and in anticipation of stringent guidelines for its EIA, environmental staff convinced senior corporate officials to build environmental considerations into the site assessment.

This was not well received by some transmission designers, who failed to understand the integrated site-selection process and were concerned that environmental staff were trying to take over their responsibility for site selection. Despite this, the importance of the project and its anticipated vulnerability in the environmental-review process enabled the environmental-integration exercise to progress, albeit with concerns. Here again, the immediate and heavy demands for completion of the site-selection work meant that the groups were unable to develop a mutually acceptable approach to the integrated site selection in a highly cooperative and orderly manner.

Although the Conawapa project and its transmission lines were later canceled and the site assessment was never fully completed, considerable progress was made though this exercise in clarifying the methodology and introducing the staff of the transmission design group to the process of integrating environmental protection into site selection. Eventually, integrated site selection became a standard part of the environmental assessment of high-voltage transmission-line projects. An acceptable working relationship and distribution of responsibilities related to site selection were established between the transmission design group and the environmental staff.

Top-level commitment to environmental management

Top-level commitment was one of several key factors that facilitated the development of five other integration tools that were voluntary initiatives of the corporation, going beyond regulatory requirements. An understanding on the part of key staff that senior executives in the corporation would support or at least not impede such initiatives facilitated development of each of these tools.

In 1989, the Board of Directors and the senior executives of Manitoba Hydro made a strong commitment to environmental management and sustainable development and encouraged corporate staff to pursue initiatives consistent with this commitment. The recently appointed chairperson of the Board of Directors, a biologist by training and an advocate of sustainable development, provided leadership in shaping this commitment. The strength and timing of the commitment were influenced by the corporation's interest in developing the new Conawapa generating station and transmission-line complex.

This project was very unpopular with Manitoba's environmental community, which was concerned about the project being developed mainly to provide power to Ontario, about flooding from the generating station, and about the prospect of transmission lines passing through undeveloped areas on the east side of Lake Winnipeg. The large scale of the project, the high level of concern in the environmental community, and the recent legal decisions broadening the role of the federal environmental-review process led to the project being designated for a federal–provincial environmental review. Such reviews tend to be wider in scope and more stringent and to involve more extensive public involvement than their provincial counterparts. Senior corporate officials recognized that during the Conawapa review it would be important for the review panel and other influential groups to perceive Manitoba Hydro as environmentally responsible.

In a large organization like Manitoba Hydro, general commitments made by senior management tend to have limited impact unless accompanied by tangible action. Such action confirms that the commitment is serious and provides leverage for groups in the organization, such as environmental staff, to push for measures to advance that commitment. During the 2 years following their decision to start Conawapa, the Board of Directors and senior executives of Manitoba Hydro demonstrated the strength of their commitment to environmental management and sustainable development by promoting and supporting three key new initiatives:

+ Launching a demand-side management program to encourage customers to make more efficient use of electricity;

+ Agreeing to settle the outstanding grievances of five Aboriginal communities for damages caused by the Grand Rapids project, although the corporation was not legally required to do so; and

+ Creating the position of Vice-President of Environmental Affairs and enlarging the group responsible for dealing with environmental matters.

These initiatives served as both a powerful signal to corporate employees to consider the environment seriously in their work and evidence that proposals for new environmental-management tools would be supported at the executive level. The new vice-presidency and the increased size of the environmental group also significantly strengthened the environmental-management capability of the organization. The vice-presidency raised the status of environmental management at Manitoba Hydro and ensured that the environmental point of view would be represented on the corporation's most important decision-making body, the Executive Committee. Creating a larger environmental unit meant that more work could be done to facilitate and promote environmentally sound decisions and actions.

Knowing that senior corporate officials were committed to environmental management made it easier for environmental staff to convince line-department staff to cooperate in the development of new environmental-integration tools. It also spurred environmental and line-department staff to extend environmental management beyond the confines of regulatory requirements into the realm of environmental stewardship. This contributed to the development of many new programs for good environmental management that were not required by law. These programs included the environmental-integration tools noted in Table 5: self-directed EIA of small projects, environmental-protection plans, the forest-enhancement program, the northern line-maintenance guidelines, and environmental costing. Without high-level commitment to environmental management, most of these tools would probably not have been developed. Staff would have probably confined themselves to developing integration tools required to satisfy regulatory requirements. Similarly, the amount of attention and effort focused on environmental management would have been reduced because the influence and size of the environmental group would probably have been smaller.

Environmental specialists on staff

Regulatory requirements and senior-level commitments provided an overall context for fostering environmental integration at Manitoba Hydro. However, priority areas for integration had to be identified and specific integration tools developed and implemented. It was expected that each department and employee would be responsible for integrating environmental protection into their activities and assist in developing suitable tools, but line-department staff lacked either the motivation and conviction or the appropriate knowledge to assume this responsibility. Using an approach employed throughout the electrical industry, Manitoba Hydro established separate organizational units, with staff appropriately qualified to promote, facilitate, and assist line departments in reducing the environmental impacts of their activities. As noted, three units — one responsible for dealing with the management of hazardous materials; another, with mitigation of and compensation for damage from past projects; and a third, with all other environmental issues — provided day-to-day leadership in furthering the corporation's environmental agenda. The size of these units almost tripled in the late 1980s and early 1990s, when environmental management was becoming a corporate priority.

As shown in the final column of Table 5, Manitoba Hydro's environmental staff played a role in developing all of these transmission-line environmental-integration tools and were the prime catalysts in developing five: EIA for large projects, self-directed EIA for small projects, integrated site selection, environmental-protection plans, and the forest-enhancement program. Environmental specialists were instrumental in ensuring that environmental management was given priority attention and in providing leadership in developing appropriate integration tools. They provided specialized knowledge in shaping the content of the tools and either played a lead role or assisted in their implementation. Although environmental specialists had a lead role in developing these integration tools, their role varied in the implementation. For example, in implementing EIAs for large projects, environmental staff spearheaded and coordinated the entire process, including dealing with regulatory agencies, working with relevant line departments, leading the public-consultation process, managing consultants to produce technical reports, and, when necessary, providing testimony at public hearings. In contrast, for the forest-enhancement program, the Construction Division had the lead role in implementation, and the environmental unit's role was to participate as a member of an interdepartmental steering committee for the program.

Environmental staff were even less prominent in the development of the northern line-maintenance guidelines and environmental costing (Table 5), largely as a result of a positive shift within the corporation. After several years, the commitment to environmental management established by senior officials and promoted by the environmental staff began to spread, as it became apparent that environmental integration was generally worthwhile and could be done without excessively disrupting mainline functions. Senior staff in many parts of Manitoba Hydro started to be more willing to incorporate environmental considerations into their activities. Some line-department staff even became proactive in the area of environmental management and started their own initiatives. This was the case with the northern line-maintenance guidelines and environmental costing, which originated from northern line-maintenance and system-planning units, rather than the environmental unit. The environmental staff found themselves in the unusual position of being asked to help develop a tool, rather than having to spearhead the process. For these tools, the main role of environmental staff was to provide specialized knowledge to help shape the tools.

Three of the seven tools — the forest-enhancement program, the northern line-maintenance guidelines, and environmental costing — were developed through interdepartmental-committee processes (Table 5). Environmental staff generally had more experience with this type of program-development committee than most others in the corporation and provided organizational and facilitative skills that enabled the committees to operate more smoothly and productively.

Getting key line staff to buy into the tools

The tools listed in Table 5 were introduced by either being imposed on the affected line departments or by persuading key staff in the affected departments to buy into them. EIA for large projects and integrated site selection were imposed; both were driven by actual or anticipated regulatory requirements and the need to meet tight schedules. In general, if regulatory requirements were involved, there was a tendency for the tools to be imposed. Environmental staff would identify a needed tool and, with the help of environmental consultants, design and shape it. The affected line departments would then be instructed to implement the tool to meet regulatory requirements. Because there was little time to explain the reasons for developing the tool or to allow staff to understand and be comfortable with it, considerable friction typically resulted between environmental and line-department staff in implementing the tools. This slowed down acceptance and reduced the sense of ownership of the tool in the line department. Initially, only the regulatory requirements kept line departments involved. However, after being exposed to a tool on several projects, key staff in the affected line departments acquired an understanding of the reasoning behind its design and were able to incorporate modifications that made them more comfortable with the way it worked for their activities. Consequently, staff in the affected departments started to take more of an interest and to develop a sense of ownership of these tools. No longer resentful of having to apply the tools, the staff started to look for ways of implementing them more effectively. Basically, they bought into these environmental-integration tools. Although making key staff buy into the tools in this way was probably necessary, given the initial resistance to environmental management, this forced buy-in was clearly less desirable than a voluntary one.

A voluntary buy-in accompanied the development of other tools. Not subject to tight schedules and with no regulatory requirements to implement them, the main pressure to develop self-directed EIA for small projects, environmental-protection plans, the northern line-maintenance guidelines, and environmental costing (Table 5) came from the commitment to environmental management at the senior level and from the high level of public interest in the environment. In these situations, where considerable discretion was available, the support of affected line departments was essential in developing and implementing a new tool. Imposing the tools would have very little likelihood of success. A voluntary buy-in was achieved by having key staff of the affected line departments involved in designing the tools and assuming the lead role in their implementation.

As noted, the buy-in processes for the northern line-maintenance guidelines and environmental costing were initiated by the line departments, whereas environmental staff initiated a process for the remainder. These tools were developed either by creating an interdepartmental committee for this purpose or through close, ongoing interaction between environmental staff and appropriate staff in the relevant line departments. This approach allowed line-department staff to present their perspectives and concerns about the scope and content of environmental-management and -integration tools for their areas and to influence the structure of the tools in relation to their activities. Trade-offs between environmental and other considerations were identified and examined, and acceptable compromises were established.

With a buy-in approach, more time was required for developing a tool than would have been required by simply imposing it, but it provided valuable dividends at the end of the process. Once the tool was approved, the affected line departments were willing to implement it, had a sense of ownership, and required little coercion by environmental staff to apply it. A great deal of acrimony between environmental and line-department staff was avoided, and generally it was easier to introduce new tools. The success with these and other buy-in processes was instrumental in setting the stage for the most ambitious and comprehensive environmental-management program ever undertaken by Manitoba Hydro — developing environmental codes of practice for all activities that affect the environment. Instigated in 1994, this initiative combined the approaches of an interdepartmental committee and, working closely with affected staff, produced more than 20 different codes of practice.

In sum, whereas the process of environmental integration initially relied on nudges from new government regulations and on senior-management commitments, it eventually became self-initiating and self-maintaining as staff in line departments became more knowledgeable and comfortable with the content and operational implications of integration tools. Voluntary buy-ins replaced prescriptions. Corporate environmental staff ensured that the difficult transition from introducing a new theme into the organization to making it an integral part of operations was successful.

Information for developing environmental-management tools and policies

As environmental management spread in Manitoba Hydro, the corporation increasingly relied on interdepartmental committees to develop environmental-management methods and policies. This proved to be an effective technique for developing workable methods and acceptable policies, which balanced the need to produce and deliver electricity cost-effectively with the need to protect the environment. This section examines the purposes, types, and sources of information used by four of these committees. The committees dealt with a wide range of circumstances, including

+ Making recommendations pertinent to planning, construction, and maintenance activities;

+ Developing both general policy and very specific integration tools;

+ Producing new environmental-management tools;

+ Developing the corporation's sustainable-development policy; and

+ Dealing with a cross section of environmental-management topics at Manitoba Hydro.

Each committee's recommendations were approved by the corporation's Executive Committee. I served on each of these committees and have drawn on my knowledge of and experience with their workings to prepare this section.

The committees were

+ *Conawapa Vegetation Management Committee* — This six-member committee clarified issues and developed recommendations for reforestation, right-of-way width, and control of forest-cover growth for the proposed Conawapa generating station, on the Nelson River, and for the transmission lines that would link the generating station to southern Manitoba. Recommendations of the committee were accepted by the Executive Committee, but they were never implemented because the Conawapa project was canceled. However, this work led to the creation of two other committees that looked into developing a corporation-wide program to offset forest-cover losses caused by corporate activities. This eventually led to the adoption of the forest-enhancement program, which includes provision for tree planting, innovative forestry, and public forestry education projects.

+ *Environmental Costing Task Force* — This 11-member committee clarified issues and developed recommendations for valuating and incorporating environmental mitigation, compensation, and externality costs into system and project planning. At the time of the research, work was under way to develop methods for valuating the environmental impacts of generating stations and transmission lines.

+ *Northern Line Maintenance Strategy Task Force* — This five-member committee was established to develop strategies to improve all aspects of transmission-line and right-of-way maintenance in northern Manitoba to cope with anticipated growth in the workload and to respond to demands for greater environmental responsibility. Among other things, the committee clarified issues and developed recommendations for incorporating environmental considerations into northern line maintenance. Its work included the development of environmental guidelines for line-maintenance staff.

+ *Environmental Working Group* — This 10-member committee was established to provide an interdepartmental forum to test and review new environmental-management ideas and to help develop a sustainable-development policy for the corporation. The committee's efforts resulted in a sustainable-development policy and 14 accompanying principles, dealing with environmental management, economic contribution, and economy–environment integration. Produced in 1993, this policy served to confirm the corporation's efforts during the preceding 5 years to become environmentally sound, rather than to chart a new direction in sustainable development and environmental management.

Information about the products, follow-up activities, and membership of each committee is presented in Tables 6 and 7.

Committees' deliberation processes

Despite having differing mandates and durations, the committees generally followed a similar process in their deliberations and made use of similar types of information. Their process followed a corporate norm that had evolved for interdepartmental committees mandated to develop new programs and policies. Although the steps in the process were not explicitly presented at the beginning of any committee's work, at least some members had previous experience with interdepartmental-committee work and brought this knowledge to the group.

The committees observed most or all of the following steps in the process in roughly the same sequence:

+ Convincing sceptical members of the committee of the need for the environmental-management tool or policy;

+ Developing a common understanding of the scope and content of activities affected by the environmental-management tool or policy and the impact of these activities on the environment;

+ Reviewing experience at Manitoba Hydro in relation to the environmental-management tool or policy;

+ Identifying options for the environmental-management tool or policy;

Table 6. Products and outcomes of corporate environmental and sustainable-development policy- and program-development committees.

	Conawapa Vegetation Management Committee	Environmental Costing Task Force	Northern Line Maintenance Strategy Task Force	Environmental Working Group
Product generated	Report recommending reforestation program for Conawapa project, to offset any losses of forest cover; recommendations for right-of-way width, control of forest growth	Report recommending that environmental mitigation and compensation costs be factored into all planning-cost estimates and that external environmental cost be considered in sensitivity analysis of planning costs	Report recommending various changes in the way northern line maintenance is carried out, including implementation of a set of environmental guidelines presented in the report	Proposed sustainable-development policy and principles for Manitoba Hydro
Follow-up outcomes	Two subsequent committees examined the feasibility, scope, and content of a reforestation program and recommended that the program apply to all proposed projects requiring environmental licences and that it become a forest-enhancement program, also covering public forestry education and innovative forestry projects. Hydro Executive approved implementation of the forest-enhancement program. Program was being implemented	Executive Committee approved recommendations; work was initiated to produce detailed methodologies for high-priority costing categories	Environmental recommendations adopted by the division manager in charge. Handbook containing environmental guidelines was prepared and supplied to each northern line-maintenance employee. Training session was held to help employees use the guidelines	Executive Committee adopted policy and principles. A set of comprehensive, more detailed environmental policies was being developed to support sustainable-development policy and principles. Sustainable-development report was produced, showing how Manitoba Hydro is implementing the sustainable-development policy and principles. Report was distributed to the public

Table 7. Actors involved with corporate environmental and sustainable-development policy- and program-development committees.

	Conawapa Vegetation Management Committee	Environmental Costing Task Force	Northern Line Maintenance Strategy Task Force	Environmental Working Group
Corporate groups shaping the committee and its work				
Group initiating the committee	Conawapa Environmental Licensing Division	Planning Division / Environmental Management Division / Financial and Economic Analysis Division	Northern Region Division	Environmental Management Division
Group chairing the committee	Conawapa Environmental Licensing Division	Generation Planning Department / Economic Analysis Department	Line Maintenance Section — Northern Region	Environmental Policy and Planning Department
Group providing secretariat support	Environmental Policy and Planning Department	Economic Analysis Department	Consultant-technical writer familiar with transmission functions	Environmental Policy and Planning Department
Membership				
Line divisions and departments with activities potentially affected by committee recommendations	Transmission and Stations Design Department / Conawapa Project Construction Department / Line Maintenance Department — Eastern Region	Power Resource Planning Department / Major Transmission Planning Department / Transmission and Stations Design Department / Hydro Development Planning Department	Line Maintenance Section — Northern Region	Planning divisions / Construction division / Production divisions / Regional divisions / Finance divisions / Personnel divisions / (One representative for each group of divisions)
Divisions and departments with an interest in producing environmental improvements	Conawapa Environmental Licensing Division / Environmental Policy and Planning Department	Environmental Policy and Planning Department	Environmental Policy and Planning Department	Environmental Management Division / Occupational Health and Safety Division
Departments with specialized knowledge to contribute on anticipated issues	Regional Services Department (company forester)	Economic Analysis Department (company economists) / Mitigation and Compensation Department (experience with compensation settlements)	Human Resource Planning and Development (training specialist) / Transmission Engineering Services / Transmission Construction	Law Department (company lawyers)

Source: InterGroup Consultants Ltd.

+ Clarifying the scope, content, costs, and effectiveness of various options and the implications for existing activities;

+ Comparing and evaluating options;

+ Reaching agreement on and refining descriptions of the preferred environmental-management tool or policy;

+ Preparing a committee report, with recommendations;

+ Preparing a submission to the Executive Committee; and

+ Receiving the decision of the Executive Committee on the recommended environmental-management tool or policy.

Types of information

Information played a role in every step of the process. In the early steps, mainly new information was used, but for later steps, the information was more a synthesis of what had already been introduced. The information differed between committees and was heavily influenced by the subject matter of their respective work. Because this work often infringed on other duties considered more important by many members, the committees wanted to proceed as quickly and efficiently as possible and place minimum demands on time for committee matters outside of the scheduled meetings. Therefore, committees were only interested in information directly related to their efforts and made no attempt to gain a broader perspective on the issues. They concentrated on solving the specific environmental-management issue they had been asked to address. Consequently, they did not take into consideration any material that presented a wider perspective, offered a conceptual approach to environmental management, or was apparently irrelevant or difficult to absorb. The committees relied almost entirely on information that was

+ Specific to the subject at hand;

+ Pertinent to the situation within Manitoba Hydro, the experience of the electricity industry, or the expectations of stakeholders of concern to Manitoba Hydro;

+ Up to date; and

+ Presented in small packages of usually no more than several pages.

Although each committee's information varied, the committees generally each used the following seven types of information:

+ Corporate commitments, policies, and experience of Manitoba Hydro;

+ Requirements or expectations of environmental regulators or key stakeholders, such as the provincial government;

✦ Descriptions of the situation, standards, practices, and procedures of the corporate activities that would be affected by the tool or policy under consideration;

✦ Environmental impacts of the affected corporate activities;

✦ Relevant emerging trends and experience of other electric utilities;

✦ Relevant design or costing data; and

✦ Materials produced by the committee during earlier steps in the process and containing a synthesis of the above types of information.

Table 8 shows which types apply specifically to each step in the committees' deliberations.

Each committee used at least six of the seven types of information noted above and little else. The depth of information varied, with the least detailed being used by the Environmental Working Group in developing the sustainable-development policy and principles; the most detailed, by the Environmental Costing Task Force, which spent almost 2 years examining procedures for integrating different categories of environmental costs into sequence and project planning.

Sources of information

The committees' information had five sources:

✦ *Knowledge of the committee members* — The committees included representatives of all potentially affected line departments, the environmental division, and departments with specialized knowledge relevant to a committee's area of work. These representatives possessed a great deal of essential information, which was usually presented verbally at meetings and recorded in the committee's meeting notes.

✦ *Invited guests* — Occasionally, people from Manitoba Hydro or other utilities known to have specialized information of interest to a committee were invited as guests to a committee meeting. They would typically make a brief presentation, followed by a discussion with members, and this was recorded in the committee's meeting notes.

✦ *Research conducted by a committee member* — The committee would often identify particular information that it needed for clarifying an issue or advancing its deliberations. The committee member best suited to obtain and analyze this information was assigned this task and asked to give a report at the next meeting, and this report usually took the form of a memo or a short working paper.

✦ *Existing documents produced by Manitoba Hydro, government, or other utilities* — Documents produced by Manitoba Hydro, the provincial

Table 8. Types of information used by Manitoba Hydro committees involved in developing environmental-management tools and policies.

Steps in committee deliberations	Types of information used
Convincing sceptical members of the committee of the need for the environmental tool or policy	Corporate environmental commitments and policies Current or anticipated regulatory requirements Trends in the Canadian–North American electrical industry
Developing a common understanding of the scope and content of activities affected by the environmental-management tool or policy and the impact of these activities on the environment	Description of setting, standards, practices, and procedures of the target activities Description of environmental impacts influenced or directly affected by target activities
Reviewing experience at Manitoba Hydro in relation to the environmental-management tool or policy	Description of relevant initiatives
Identifying options for the environmental-management tool or policy	Description of how target activities contribute to environmental impacts Approaches required or advocated by regulatory agencies and other key stakeholders, such as the provincial government Approaches used by other electric utilities in similar situations Opportunities and problems associated with applying potential approaches, drawing heavily on information about how existing target activities are performed
Clarifying the scope, content, costs, and effectiveness of proposed options and the implications for existing activities	Option-specific design and costing data Detailed descriptions of comparable initiatives adopted by other electric utilities Description of setting, standards, practices, and procedures of the target activities
Comparing and evaluating options	Material produced in the preceding step
Reaching agreement on and refining descriptions of the preferred environmental-management tool or policy	Option-specific design and costing data
Preparing committee report with recommendations	Material produced throughout committee deliberations
Preparing a submission to the Executive Committee	Material produced in committee report
Receiving the decision of the Executive Committee on the recommended environmental-management tool or policy	Executive submission, committee report, and comments solicited from staff

Source: InterGroup Consultants Ltd.

or federal government, and other utilities often contained information important to a committee's deliberations. Applicable portions of these documents would be distributed to all committee members.

+ *Professional research articles or studies* — Occasionally, a committee member would have access to a useful study or article produced by an academic institute or research organization. Relevant portions would be distributed to all committee members.

Information supplied to or used by each committee is shown in four tables: Table 9 covers knowledge of committee members; Table 10, knowledge of invited guests; Table 11, research conducted by committee members; and Table 12, government and other-utility documents. A consolidation of these listings is presented in Table 13, which shows the relationship between the types and sources of information and indicates how the four committees made use of each source. The following perspectives on the sources of information emerged from this consolidation:

+ The most frequently used source of information was the knowledge of committee members. Every committee used this as their primary source for topics specific to Manitoba Hydro and its activities. Committee members also commonly supplied information about regulatory requirements and stakeholder expectations.

+ The sources of information used for most purposes were research conducted by committee members and existing documents from Manitoba Hydro, government, or other utilities. Research conducted by committee members was used to supply five types of information (Table 13) and was frequently used to provide information on emerging trends and experience of other electric utilities, as well as option-specific design and costing data. Existing documents were also used to provide five types of information but not frequently applied in any single category. This was the only source used by more than one committee in providing option-specific design and costing data.

+ The least used sources were invited guests and professional research articles and studies. These were used to provide two types of information (Table 13) and were infrequently used by any committee.

The above findings are confirmed by my own impressions of the relative importance of the five sources of information. Committee members' knowledge and the research they conducted played a vital role in the deliberations of all four committees. Existing documents from Manitoba Hydro, government, or other utilities were very important to the committee developing the sustainable-development policy and principles but of minimal to moderate importance to the work of two of the other committees. Invited guests and professional research articles and studies were occasionally used and tended to have a minor influence.

Table 9. Knowledge supplied by members of corporate environmental and sustainable-development policy- and program-development committees.

	Conawapa Vegetation Management Committee	Environmental Costing Task Force	Northern Line Maintenance Strategy Task Force	Environmental Working Group
Affected line-department representatives	Current practices and their underlying rationale for determining right-of-way widths, clearing of generating station forebays and transmission lines, vegetation-control methods for transmission lines; contacts in other utilities	Current practices and their underlying rationale for estimating costs used in planning sequence analysis, next-plant planning, generating-station design, transmission location and design	Current practices and their underlying rationale for northern line inspection and repair and for vegetation control; way in which line-maintenance activities directly affect the environment	General understanding of current practices for the planning, design, construction, operation, maintenance, financial, and personnel functions of the corporation In the case of two representatives, familiarity with *Our Common Future* (WCED 1988) and key features of sustainable development
Environmental representatives	Emerging trends in regulatory requirements and public concerns related to vegetation management; specific needs of Conawapa environmental assessment	Conceptual framework for environmental costing; relationship between EIA and environmental costing	Current and emerging environmental-management policies of the corporation; appreciation of the main environmental concerns associated with transmission lines	Understanding of the main features of sustainable development and environmental management; familiarity with Manitoba government's principles of sustainable development; knowledge of corporate initiatives relevant to sustainable development
Representatives of specialist departments	Impacts that clearing for generation and transmission would have on forest cover in areas affected by the Conawapa project; contacts in other utilities and forestry agencies	Corporate experience with compensation settlements; conceptual framework for environmental costing	Experience with applying environmental-protection measures in construction activities	Legislative context for environmental management

Source: InterGroup Consultants Ltd.
Note: EIA, environmental-impact assessment.

Table 10. Knowledge supplied by invited guests to corporate environmental and sustainable-development policy- and program-development committees.

	Environmental Costing Task Force	Northern Line Maintenance Strategy Task Force
Presentations by or discussions with invited guests, or both	Presentation by a consulting economist from Toronto about work he had done for Ontario Hydro on integration of environmental externalities	Discussion in Thompson with the entire line-maintenance staff about working conditions during line inspections in winter and their perspectives on protecting the environment while on the job Presentation by manager of line maintenance for Ontario Hydro about its line-maintenance practices, including environmental-protection measures Presentation by three members of environmental-management staff about environmental issues related to line maintenance they were familiar with, including those raised in the course of public-consultation exercises for transmission-line EIA, such as use of herbicides for vegetation control
Basic attributes of information used	Current information specific to North American electrical industry and topics being examined by the committee	Current information specific to Manitoba Hydro and topics being examined by the committee
Reasons for using information	To clarify trends emerging in the electrical industry on the topic being addressed by the committee and generate practical ideas for addressing the topics being considered by the committee	To clarify trends emerging in Manitoba on the topic being addressed by the committee and generate practical ideas for addressing the topics being considered by the committee
Importance to process and product	Minimally important	Moderately important

Source: InterGroup Consultants Ltd.
Note: EIA, environmental-impact assessment.

Table 11. Research by members of corporate environmental and sustainable-development policy- and program-development committees.

	Conawapa Vegetation Management Committee	Environmental Costing Task Force	Northern Line Maintenance Strategy Task Force	Environmental Working Group
Working papers or memoranda about relevant existing policies and practices at Manitoba Hydro (usually produced by secretariat or appropriately qualified committee member in response to committee request; supplied in advance of meeting and presented at meeting by the person who prepared document)	Review of Manitoba Hydro's current transmission-line clearing standards Review of existing Manitoba Hydro documentation on vegetation control Summary of herbicide use by Manitoba Hydro for vegetation control	Review of main environmental impacts of Manitoba Hydro's activities and the relevant information available at Manitoba Hydro for estimating their environmental costs		Compendium of existing corporate policies and senior-management pronouncements related to environmental management and sustainable development
Basic attributes of information used	Current information specific to Manitoba Hydro and topics being examined by the committee	Current information specific to Manitoba Hydro and topics being examined by the committee		Current information specific to Manitoba Hydro and topics being examined by the committee
Reasons for using information	To clarify the corporate setting for committee deliberations and discussions	To clarify the corporate setting for committee deliberations and discussions		To clarify the corporate setting for committee deliberations and discussions
Importance to process and product	Very important	Moderately important		Minimally important

(continued)

Table 11 concluded.

	Conawapa Vegetation Management Committee	Environmental Costing Task Force	Northern Line Maintenance Strategy Task Force	Environmental Working Group
Working papers or memoranda to better understand a specific issue or potential recommendation (usually produced by secretariat or appropriately qualified committee member in response to committee request; supplied in advance of meeting and presented at meeting by the person who prepared document)	List of potential reasons for developing a reforestation program Estimates of forest cover to be removed by Conawapa flooding and by clearing for its associated transmission lines Estimate of reforestation requirements and costs for the Conawapa project Evaluation of five reforestation alternatives for the Conawapa complex	Conceptual framework for valuation and integration of environmental costs and benefits (later presented as paper at conference on electric-utility planning) Test case for including external environmental-cost estimates of air emissions in development-sequence analysis (used estimates of unit-emissions externality costs from five recent sources) Evaluation of options for the treatment of environmental externalities by Manitoba Hydro	Draft of environmental guidelines for northern line maintenance (derived mainly from Ontario Hydro guidelines and knowledge of Manitoba Hydro environmental specialist with experience in developing environmental-protection plans for new transmission lines)	
Basic attributes of information used	Analysis and synthesis of ideas or data specific to issue being examined by the committee	Analysis and synthesis of ideas or data specific to issue being examined by the committee	Analysis and synthesis of ideas or data specific to issue being examined by the committee	
Reasons for using information	To meet a specific need identified by the committee to clarify an issue and advance its deliberations	To meet a specific need identified by the committee to clarify an issue and advance its deliberations	To meet a specific need identified by the committee to clarify an issue and advance its deliberations	
Importance to process and product	Very important	Very important	Very important	

Working papers or memoranda about relevant experience of other utilities (usually produced by secretariat or appropriately qualified committee member in response to committee request; supplied in advance of meeting and presented at meeting by the person who prepared document)	Review of recent developments in vegetation management at six other Canadian utilities, with emphasis on the corporate reforestation program that had recently been established by Ontario Hydro	Review of status of internalization of externalities in Canadian electric utilities	Review of recent developments in line maintenance (including environmental protection) in seven other Canadian utilities
Basic attributes of information used	Current information specific to electric utilities in Canada and topics being examined by the committee	Current information specific to electric utilities in Canada and topics being examined by the committee	Current information specific to electric utilities in Canada and topics being examined by the committee
Reasons for using information	To clarify trends emerging in the electrical industry on the topic being addressed by the committee; to assess how Manitoba Hydro compares with other major Canadian utilities on the topic being addressed by the committee; to generate practical ideas for addressing the topics being considered by the committee	To clarify trends emerging in the electrical industry on the topic being addressed by the committee; to assess how Manitoba Hydro compares with other major Canadian utilities on the topic being addressed by the committee; to generate practical ideas for addressing the topics being considered by the committee	To clarify trends emerging in the electrical industry on the topic being addressed by the committee; to assess how Manitoba Hydro compares with other major Canadian utilities on the topic being addressed by the committee; to generate practical ideas for addressing the topics being considered by the committee
Importance to process and product	Very important	Moderately important	Moderately important

Source: InterGroup Consultants Ltd.

Table 12. Government and other-utility documents used by corporate environmental and sustainable-development policy- and program-development committees.

	Environmental Costing Task Force	Northern Line Maintenance Strategy Task Force	Environmental Working Group
Relevant documents produced by government agencies or other utilities (usually supplied to the committee by secretariat and commented on at next meeting by the person supplying information)	External-cost estimates and studies produced by Bonneville Power, regarded as the US electric-utility leader in this field	Environmental guidelines for transmission-line construction and maintenance produced by Ontario Hydro, regarded as Canadian leader in the field	Government of Manitoba's sustainable-development principles (this was the key document to the entire exercise as it was used as the basis for Manitoba Hydro policy and principles) Memo from Manitoba Round Table on Environment and Economy to businesses in Manitoba, inviting them to commit to and develop sustainable-development policies Section of recent provincial *Mines and Minerals Act* referring to sustainable development Canadian Electric Association paper describing 11 propositions for sustainable development Canadian Electric Association statement on sustainable development Article by the President of TransAlta utilities about its interest in being a leader in sustainable development
Basic attributes of information used	Current information specific to a leading hydroelectric utility in the United States and topics being examined by the committee	Current information specific to a leading electric utility in Canada and topics being examined by the committee	Current information about Manitoba government initiatives affecting Manitoba Hydro with respect to the topics being addressed by the committee Current information specific to Canadian electric utilities and topics being examined by the committee
Reasons for using information	To generate practical ideas for addressing the topics being considered by the committee	To introduce practical ideas for addressing topics being considered by the committee	To confirm the importance of committee's work and generate ideas for addressing the topics being considered by the committee
Importance to process and product	Minimally important	Moderately important	Very important

Source: InterGroup Consultants Ltd.

Table 13. Sources of information used by four Manitoba Hydro committees involved in developing environmental-management tools and policies.

Types of information used	Sources used				
	Knowledge of committee members (usually presented verbally)	Invited guests with relevant knowledge	Research conducted by committee members (usually presented as working papers or memoranda)	Existing Manitoba Hydro, government, and other-utility documents	Professional research articles or studies
Corporate commitments, policies, and experience related to the tool or program	4 (Environmental representative)		1	1 (Manitoba Hydro documents)	
Requirements or expectations of regulators and key stakeholders	2 (Environmental representative)	1 (Corporate environmental specialists)		1 (Regulatory and provincial-government documents)	1 (Key study on sustainable development)
Descriptions of situation, standards, practices, and procedures of targeted activities	4 (Line-department representatives, also one orientation trip)		1	1 (Manitoba Hydro documents)	
Environmental impacts of target activities	3 (Line-department and environmental representatives)		1		
Emerging trends and experience of other electric utilities		2 (Consultant working for another electric utility, manager of another electric utility)	3	1	1 (Recent major electric-utility study and two recent journal articles related to the electrical industry)
Option-specific design and costing data			3	2 (Reports from other electric utilities)	

Source: InterGroup Consultants Ltd.
Note: Numbers refer to the number of committees using a particular type of information.

The role of professional research articles and studies, a source used by two committees, is shown in Table 14. The Environmental Costing Task Force used material from a comprehensive study on incorporating environmental externalities into electric utility planning, conducted by Pace University Center for Environmental Legal Studies in 1990, and from articles in two special issues of the *Electricity Journal* that dealt with incorporating environmental externalities. Both these sources provided fairly current information specific to the electrical industry on the topic addressed by the committee. Although the information clarified trends in the Canadian and US electrical industry with respect to incorporating externalities into planning, it was of minimal importance because it was superseded by more recent data on the Canadian electrical industry obtained through research by a committee member. The other research study, which was used by the Environmental Working Group, was *Our Common Future* (WCED 1988), generally regarded as the seminal document on sustainable development. An abbreviated version was provided to committee members involved in developing the corporation's sustainable-development policy and principles. Its eventual impact on committee deliberations was minimal because it was not specific enough to Manitoba Hydro's situation. Occasionally, a committee member would refer to *Our Common Future*, but this had little influence

Table 14. Research studies and journal articles used by corporate environmental and sustainable-development policy- and program-development committees.

	Environmental Costing Task Force	Environmental Working Group
Research studies (usually supplied to the committee by secretariat and commented on at next meeting by the person supplying information)	Pace University Center for Environmental Legal Studies (1990) (introductory section and section on hydroelectric valuation) CECARF (1993) (sections on current status of US state activities to incorporate externalities, current externality values used in utility-resource decision-making)	Condensed version of *Our Common Future* (WCED 1988)
Journal articles	Several articles from two special issues of the *Electricity Journal* (about incorporating environmental externalities)	
Basic attributes of information used	Current information specific to the North American electrical industry and topics being examined by the committee	
Reason for using information	To clarify trends emerging in the electrical industry on the topic being addressed by the committee and generate practical ideas for addressing the topics being addressed by the committee	Seminal study that influenced the evolution of sustainable development
Importance to process and product	Minimally important	Minimally important

Source: InterGroup Consultants Ltd.

on the principles of sustainable development adopted by the committee. The content of these principles was largely influenced by the knowledge of committee members about the Manitoba Hydro setting and the sustainable-development principles adopted by the Manitoba Round Table on Environment and the Economy.

Information not used

The information used by the committees tended to be highly focused. Certain types of information that outsiders would have expected the committees to use were overlooked (see Table 15). For example, although the International Institute for Sustainable Development is located in Winnipeg and is involved in examining environmental-integration issues, none of its reports were used. Its scope was considered too broad and focused on neither the electrical industry nor specific activities in the industry, such as transmission-line maintenance or vegetation control. Similarly, the Environmental Costing Task Force made no use of the large body of theoretical and analytical literature on environmental economics because it was too conceptual and failed to deal with practical applications in the areas being considered by the committees. The full version of *Our Common Future* was not used because the document was deemed too long and general. My impression is that a reluctance to use broad and conceptual information is not unique to Manitoba Hydro. The practical considerations driving the committees would likely be found in other large business organizations. Policy and program-development committees in other electric utilities and major industrial enterprises probably would also make use of similar types and sources of information.

Committees are one of a number of approaches that Manitoba Hydro used to develop environmental programs and policies. The other commonly used approach was to have outside consultants or staff in the environmental areas develop an initial version of the program or policy, which would then be distributed to interested departments for review and comments, with a final version taking into account the comments received. My experience with this shows that a narrower range of information is used in this approach to the developmental process than when committees are involved. Once again, the information is highly focused on the assignment at hand.

Highlights and recommendations

Environmental integration

This case study examined the process of environmental integration of transmission-line projects at Manitoba Hydro. The experience is probably quite typical of how environmental integration occurs in large, multilayered organizations engaged in natural-resource development. The following insights, gained from

Table 15. Types of information not used by corporate environmental and sustainable development policy- and program-development committees.

	Conawapa Vegetation Management Committee	Environmental Costing Task Force	Northern Line Maintenance Strategy Task Force	Environmental Working Group
Category of information	Professional literature on reforestation, vegetation control	Articles on environmental externalities from economics and environmental journals	Research reports from International Institute for Sustainable Development and related organizations about incorporating environmental considerations into business practices	Full version of Our Common Future (WCED 1988) Reports produced by the International Institute for Sustainable Development in Winnipeg
Reason for not using the information	Could not find current articles related to applications by the electrical industry	Not focused on the electrical industry; too theoretical and failed to deal with practical-application issues; typically not current enough in terms of actual developments in the area	Too general; not focused on transmission-line maintenance activities	Too lengthy; not focused on practical-application issues; not current enough Too lengthy; not focused on the electrical industry

Source: InterGroup Consultants Ltd.

the case study, may serve as a starting point for shaping the principles and goals of an approach to environmental integration in electric utilities and other large organizations operating in developing countries.

+ Environmental integration occurs when an organization's decisions and actions related to an activity continuously take into consideration the environmental consequences of that activity and incorporate measures to manage these consequences. An important goal of the environmental-integration process is to have environmental-management tools applied to all of a corporation's environment-affecting activities.

+ The process of environmental integration involves an organization's progressively introducing and reinforcing a willingness to consider environmental effects among the organizational units whose actions and decisions affect the environment. It is unlikely to occur in a planned, linear fashion but is more likely to take place incrementally in response to specific demands or opportunities, proceeding from one organizational unit to another and from one activity to another over a series of projects.

+ The environmental-integration process typically begins slowly, often requiring a push start to break down the initial apprehension about adding new, poorly understood decision-making and action criteria to existing operational systems. However, as more units in an organization gain experience with environmental integration and find acceptable ways to incorporate environmental considerations, the integration process can build up a momentum of its own and spread voluntarily throughout the organization. Some of the staff who learn about and become comfortable with factoring the environment into their decisions and actions may become advocates of environmental integration.

+ Environmental-management tools, such as EIA, environmental codes of practice, and renewable-resource replacement programs, are practical mechanisms for factoring environmental considerations into actions and decisions and give operational meaning to the concept of environmental integration. The essence of environmental integration is to identify appropriate tools for specific activities, design them to fit well with the primary corporate activity they are intended to influence, and gain acceptance from the affected organizational unit for their implementation. Environmental-management tools must be tailored to particular functions and activities if they are to be effective and accepted.

+ Regulatory requirements are the most common catalysts for initiating the environmental-integration process. Other catalysts include senior-management commitments and public pressure on the organization.

✦ The process of integration can be facilitated and expedited by having environmental specialists on staff who are responsible for establishing environmental-management priorities and for promoting and assisting with the development of environmental-management tools by divisions and departments.

✦ Environmental integration is most effective if the staff in the organizational units responsible for implementing particular environmental-management tools in one way or another buy into the tools. The most satisfactory way of getting a buy-in is to involve these staff in the planning and design of the management tools, rather than imposing the tools on them.

Based on these insights and other information gleaned from the case study, I recommend that electric utilities and other major resource-development companies in developing countries take the following steps to either introduce or accelerate environmental integration:

1. Have senior management make a commitment to being environmentally responsible. Initially, adopt a very simple position on minimizing environmental effects of corporate activities and ensure that this is made known throughout the organization. Also, make it clear that senior management is prepared to tolerate some extra cost to fulfil this commitment and that this objective will be a basis for evaluating employee performance. Senior managers must follow up on this commitment with decisions and actions to demonstrate that they are serious. A catalyst may be required for senior management to take this step, such as the appointment of a new Board of Directors or chief executive officer; public pressure with a focus on an environmental problem caused by the corporation; or the threat of tougher environmental laws or regulations.

2. Establish a small organizational unit that will be responsible for coordinating and facilitating environmental management and will report directly to senior management. The unit's first task should be to identify priority areas for environmental integration, on the basis of such factors as the seriousness of environmental damage, actual or potential regulatory noncompliance or public pressure, and openness of the organizational unit.

3. Employ staff in the environmental unit with a range of skills they can contribute to environmental integration, including specialized knowledge of environmental impacts and management tools, familiarity with the organization and its decision-making process, and experience with getting staff to buy in.

4. Give the environmental-management unit the authority to initiate processes for developing appropriate environmental-integration tools for the areas requiring priority attention. Through intensive consultation or technical working committees, ensure that staff from the responsible departments are involved from the beginning in designing the tool so an acceptable tool is developed and the affected staff have a sense of ownership of the tool during its implementation. Once priority areas have been addressed in this way, new advocates for environmental integration will likely emerge, and the process of extending integration to other parts of the organization is likely to become self-generating.

5. Once a pattern and style of environmental integration have been established in the organization, obtain a formal commitment from the environmental-management unit and senior management by having them create a reasonably detailed environmental-management policy, supported by descriptions of environmental-management tools (also known as environmental codes of practice).

If corporations are reluctant to make a strong commitment to environmental management on their own, government intervention, in the form of new or tougher environmental laws and regulations, is required to kick-start the process. Analysis suggests, however, that government does not necessarily need to be involved in regulating every corporate activity affecting the environment. It is probably sufficient to regulate corporate activities with the most serious impacts or generating the greatest public concern. Where possible, it would be desirable to phase in the regulations gradually so that the time can be taken at the corporate level to get implementing departments to properly buy into the changes.

Information types and sources

Another area examined in the study was that of information types and sources used for developing corporate environmental-management tools and policies. Four environment-related policy- and program-development exercises conducted by Manitoba Hydro committees were studied in detail; key findings from this research are the following:

- ✦ The committees engaged in these exercises concentrated on solving a particular problem or addressing a prescribed issue.
- ✦ The information the committees used was very practical and focused and possessed the following basic traits:
  - ✧ It was specific to the subject at hand;
  - ✧ It pertained to the situation within the corporation, the experience of the industry, or the expectations of relevant stakeholders, such as government and shareholders;

⋄ It was up to date; and

⋄ It was presented in small packages of no more than several pages.

✦ Very little attention was paid to information that presented a wider perspective or offered purely conceptual approaches.

✦ The main sources of information were knowledge of committee members, research conducted by committee members, and existing documents from Manitoba Hydro, government, and other utilities.

✦ Very little use was made of professional research articles and studies, which therefore had limited impact on committee deliberations.

These findings have implications for research organizations that wish to increase their funding for environmental-management research from corporate sources, such as electric utilities and other organizations engaged in natural-resource development. Most of this type of funding is likely to be tied to research exercises that are immediately relevant to the sponsoring organizations and capable of assisting them in performing their day-to-day activities. Corporate sponsors are unlikely to be interested in supporting environmental-management research that is done on the level of industrial activity in general; is conceptual, or theoretical, in nature; is concerned primarily with past experience; or is lengthy and time consuming to read. To succeed in garnering significant corporate support for environmental-management research, research organizations must be prepared to produce studies that are industry or activity specific, practical, up-to-date, and bite sized. The studies would probably have to be focused on particular countries or at least on groups of nearby countries with similar socioeconomic circumstances.

Chapter 4

Managing for Sustainability

S. Owen

◇ ◆ ◇

Introduction

People's appreciation of the importance of managing the world's resources for the benefit of future generations has grown dramatically in the last quarter century, since a photograph from space graphically illustrated the fragility of the planet. Just a decade ago, the Brundtland report (WCED 1988) emphasized the need for global action to ensure sustainable development based on a clear understanding that the sustainability of a healthy economy depends on sustaining a healthy environment.

The impact of economic activity on the sustainability of renewable resources, such as timber and fish, and on the natural environment as a whole has been an often repeated concern over many decades. However, only in recent years has there been widespread public and political resolve to take concerted action. What form that action should take remains the subject of considerable controversy.

The integration of economic, social, and environmental policy is necessary for managing for the sustainability of mutually dependent economies, communities, and environments. This integration must be achieved in the face of institutional structures and mythologies deeply rooted in the needs and assumptions predating sustainability concerns. This friction is especially evident in a country such as Canada, with its massive land base, small population, and rich supply of natural resources that have long been the envy of the world. In British Columbia in particular, the economy was built on the export of a seemingly inexhaustible supply of natural resources. That wealth of resources and wilderness is key to the mythology that shaped the identity of the province and its institutional structures.

Events of the past few years have shown us that we need to transform our approach to land and resource management and that we need to reexamine our perspective on British Columbia. But a century-old myth is not easily transformed in a few years. A fundamental challenge in learning how to manage for sustainability lies in learning how to adjust deeply rooted attitudes.

In a relatively short time, British Columbia has grown from an undeveloped frontier culture to an affluent and sophisticated society that is demonstrating leadership in management for sustainability. This chapter describes the advances that have been made and the obstacles in the way of management for sustainability. Close attention is paid to public participation (through multisector, interest-based negotiations) in preparing land-use and resource-management plans. Until recently, such negotiations were rare and were regarded by government as largely incidental to the decision-making process; now, however, government recognizes that the most effective way of achieving sustainability is through shared decision-making processes that reveal, balance, and accommodate a full spectrum of interests. Where, as in the case of sustainability, government requires a strong public commitment to implement decisions, the benefits of shared decision-making are becoming increasingly clear. Shared decision-making in the development of public policy is relatively new and is characterized by ongoing development. Much has been learned recently about the efficient practice of shared decision-making, but much remains to be learned.

British Columbia: a general description

Located on Canada's Pacific coast, British Columbia is the country's third largest province in area and in population. With a land and freshwater area of $953\,046$ km^2 (94 million ha), the province is larger than the combined area of the US states of Washington, Oregon, and California. The population of about 3.7 million is largely concentrated in a few urban areas in the southern half of the province; more than one-half of British Columbia's residents live in Vancouver and nearby communities near the mouth of the Fraser River. The remaining 1.5 million people occupy an area larger than France and Germany combined. More than 90% of this vast area is publicly owned Crown land.

The topography of British Columbia is dominated by the mountain ranges that lie on a northwest–southeast axis in the western Americas. Between the province's two major mountain systems — the Rockies to the east and the Coast Mountains to the west — lies an extensive area of rolling forest and grassland. The northeastern part of the province — to the east of the Rockies — is lowland prairie, part of the continent's Great Plains. About 1% of the province is agricultural land. The 7000-km coastline is largely uninhabited. Ice-free, deep-water ports have favoured the development of the province's large shipping industry. The waters off the mainland are sheltered by an abundance of islands;

the most significant of these are Vancouver Island and the Queen Charlotte Islands.

The varied topography and climatic patterns in the province contribute to significant ecological diversity. Six of Canada's 11 major biotic regions are represented in British Columbia, in 13 biogeoclimatic zones. The province has rainforests, subarctic areas, and deserts and has more natural diversity than any state or province in North America.

Most of the province is covered in coniferous forest, including Canada's wettest and driest forests. These publicly owned forests provide the timber that for more than a century has been a driving force in the provincial economy, and many of the province's rural communities remain almost wholly dependent on the availability of timber supplies. Forest companies hold the right to cut timber, primarily through renewable licences issued by the provincial Ministry of Forests. Other major sources of resource-based revenue include mining, oil and gas, agriculture, fisheries, and hydroelectric power. Particularly in the last few decades, tourism has played an increasingly prominent role in the provincial economy.

Historical background

First Nations

The Aboriginal peoples who have occupied British Columbia for several thousand years had developed sophisticated cultures by the time European explorers arrived on the Pacific coast, in the latter part of the 18th century. European interest initially focused on the trade in furs, leading to the virtual extinction of the sea otter. In 1858, when the discovery of gold in the Cariboo country led to a rush of prospectors from the United States, mainland British Columbia was declared a British colony.

Both by accident and by design, contact with the newcomers was devastating for Aboriginal people. The Aboriginal population was reduced to a fraction of its former size by epidemics of smallpox and other diseases introduced by European sailors. Aboriginal lands were appropriated by the British government, generally without treaty, and the people were confined to small reserves. Later, the Canadian government instituted policies that had the effect of eroding Aboriginal cultures; these policies included the prohibition of the potlatch ceremony and enforced placement of Aboriginal children in residential schools, where their languages were banned. Since that time Canada's Aboriginal peoples have become increasingly successful in making the resolution of these long-standing injustices a priority for action by the provincial and federal governments.

Economic growth

Vancouver today is a dynamic, cosmopolitan city, with a prominent role in trade among Pacific Rim countries; yet it is one of the youngest major cities on the continent. It was incorporated only 110 years ago, and a decade before that was virtually unpopulated rainforest.

When British Columbia joined the Canadian Confederation in 1871, many eastern Canadians were sceptical about the value of building a 4 827-km railway to link the province to the rest of the country, dismissing British Columbia as a "sea of mountains." However, the completion of the Canadian Pacific Railway, in 1885, provided the catalyst that launched the province's timber industry. Urban growth in various parts of the world created a sudden and heavy demand for the seemingly inexhaustible supplies of British Columbia timber and especially for straight-grained Douglas fir. Mining was the other primary contributor to the economy in the early years of the province.

Sustainability of timber supplies has been a concern in British Columbia for decades, although the issue has captured widespread public attention only in recent years. A 1945 Royal Commission on the forest industry noted that

> it is not too late to plan now for the future, but the sands are running out and the time is now upon us when the present policy of unmanaged liquidation of our forest wealth must give way to the imperative concept of a planned forest policy designed to maintain our forests upon the principle of sustained-yield production.
>
> Sloan (1945, pp. 9–10)

The government adopted the commission's recommendation that it establish long-term tenures to provide the incentive for careful management. In later years, however, technological advances and the consolidation of the industry into a few large companies contributed to a 400% increase in the annual cut between 1956 and 1976, accompanied by unprecedented prosperity in the province. Although government and industry alike subscribed to the principle of sustainability, the rewards of economic growth, combined with widespread public belief in the abundance of British Columbia's resource wealth, acted as a disincentive to concerted action.

Land-use conflict

Growth and prosperity also had consequences that set the stage for later conflict over the use of British Columbia's public lands. For example, increased demand for recreational wilderness experiences, combined with the increasing scarcity of wilderness, led advocacy groups to lobby for the protection of areas where outdoor recreation could be enjoyed. In later years, growing public concern about the impact of human development on the environment added another element to the debate about the management of public lands. Groups in the environmental movement have a variety of goals, such as protection of water quality, expansion of protected areas, and preservation of intact ecosystems. The result has been a gradual increase in the number of vocal and potentially conflicting interests in the use of public, especially forested lands.

The 1980s was a decade of intense land-use conflicts in British Columbia. The Brundtland report (WCED 1988), with its emphasis on the urgent need for sustainable development in resource economies throughout the world, sharpened the focus of the conflict. The rapidly growing tourism industry's need for unspoiled landscapes to attract its clientele was frequently at odds with the requirements of the forest industry for fresh timber supplies in pristine areas. The British Columbia land-use debate was further complicated by the commitment of the provincial, federal, and First Nations governments to resolve outstanding land, resource-ownership, and jurisdictional issues through negotiation of treaties.

Government attempts to manage the growing conflict were hampered by an organizational structure that remained oriented almost exclusively to the economic needs of British Columbia's resource economy. The Ministry of Forests retained administrative authority over the vast majority of public lands by virtue of their being designated "provincial forest." By comparison, the new ministries of environment, tourism, and recreation had little influence, and coordination of efforts among ministries was extremely limited. The Environment and Land Use Committee secretariat, established in the early 1970s to resolve land-use issues involving conflicting government agencies, was disbanded in 1980, as the province was struggling to deal with an economic recession.

Parties to the conflict who were not directly associated with resource industries complained that they were excluded from a decision-making process that favoured resource development. Their complaints were frequently addressed at the Ministry of Forests, which had the dual and, in many eyes, conflicting responsibility for maximizing the economic benefit of the timber-harvesting industry while ensuring the recognition and accommodation of other values in the forests — more than 80% of all land in the province was legally designated as provincial forest and hence came under the management of the ministry. As a means of influencing the decision-making process, environmental and First Nations groups resorted to dramatic, publicity-gaining activities, including road blockades, leading to mass arrests in widely separated areas such as the Slocan Valley, the Queen Charlotte Islands, and western Vancouver Island.

Moving beyond polarization

The events of the 1980s led many British Columbians to reexamine the long-prevailing perception of their province as a wilderness with a rich and essentially limitless bounty of natural resources and unspoiled landscapes. Conflict among people with competing values led to a growing appreciation of the need to ensure — through careful planning — the sustainability not only of timber supplies but also of other resources, such as fisheries, and nonextractive natural resources, such as biological diversity, wilderness, and landscapes. With this appreciation came a deeper understanding of the interdependent nature of the needs and values that had previously been viewed as separate and in competition; for example, British Columbia's communities, both rural and urban, depend on a healthy resource-based economy, and the future of renewable resources, such as timber and fish, depends on the maintenance of healthy, diverse ecosystems. It was also apparent that British Columbia had a luxury unavailable in many other parts of the world — the combination of a relatively small population, a large and publicly owned land base, and a sophisticated infrastructure gave it the opportunity to plan for a sustainable future and to set a world standard. All that was needed was the political and public will to take the risk, recognizing that the rewards of doing so could far outweigh the pain of the required economic and social adjustments.

The costs of conflict, together with the growing recognition of the need to address sustainability issues, led to calls in the late 1980s and early 1990s for a comprehensive and coordinated government strategy for the management of public lands and resources. In 1992 the government set up the Commission on Resources and Environment (CORE), with a statutory mandate to advise the government on land-use, resource, and environmental issues and to develop a comprehensive sustainable-land-use strategy, including regional planning and broad public participation in decision-making.

Commission mandate and activities

The *Commissioner on Resources and Environment Act* (GBC 1992), passed by the British Columbia legislature in July 1992, states that CORE's primary responsibility is to "develop for public and government consideration a British Columbia wide strategy for land use and related resource and environmental management." The statute provided that, in the preparation of this provincial land-use strategy, CORE was to undertake the following related tasks:

+ Establish strategic provincial direction to develop new policy and coordinate and integrate existing policies — Some policies were already in a state of rapid transition as provincial agencies attempted to meet public expectations for increased integration and efficiency, as well as for more sustainable resource uses.

+ Develop and monitor regional land-use plans to resolve broad land-allocation issues, particularly those related to the creation of new protected areas, in keeping with a government commitment to increase the amount of protected land in the province from 6% to 12% — Allocation conflicts had escalated to the point where civil disobedience, including road blockades and tree spiking, and sophisticated public information and lobbying campaigns had become common, effectively destabilizing decision-making processes and increasing uncertainty for all concerned. Previously, these issues had been dealt with on an ad hoc, valley-by-valley basis. Regional plans would allow them to be dealt with in a more strategic and efficient manner.

✦ Design a more effective means for communities (including the full range of resource and environmental interests) to participate in resource-management decision-making at the local level on an ongoing basis — A wide variety of community-based organizations (round tables, resource associations, etc.) had already been established across the province as citizens attempted to find new ways to participate in government decision-making while reconciling differences among themselves. These initiatives were driven by a general dissatisfaction with the conventional public-consultation processes, which made delegated decision-makers the final arbiters of conflicting public interests and gave limited opportunity for opposing interests to reconcile their differences themselves.

✦ Design a dispute-resolution system for land-use and resource and environmental issues, with emphasis on dispute prevention, as well as on ensuring administrative fairness and accountability.

✦ Ensure that Aboriginal rights and treaty negotiations would be respected and not prejudiced; that is, if First Nations chose to participate in developing the land-use strategy, the resulting decisions would not be binding on them in subsequent treaty negotiations — The province had recently reversed a 100-year-old policy of not negotiating treaties with First Nations and was now committed to resolving outstanding land claims through treaty negotiations. Generally, people recognized that provincial lands and resources affected by a land-use strategy might also be subject to treaty agreements. People also recognized that the resolution of land-use issues was required sooner than treaties could be negotiated but also that a land-use strategy oriented to sustainability would be in everyone's best interests, regardless of the outcome of negotiations on land ownership and jurisdiction.

✦ Provide opportunities for economic, environmental, and social interests, as well as local, provincial, and federal governments and Aboriginal peoples, to participate in the development and implementation of the land-use strategy.

✦ Oversee and monitor the ongoing implementation of the provincial land-use strategy.

A chronological progression from provincial-policy development, to land-use planning, to local implementation and monitoring would have been the orderly approach to developing the strategy, but the intensity of the issues and the widespread public expectation of action forced CORE to proceed on all aspects of its mandate at once. The urgency of the situation also compelled government ministries to proceed independently in developing key policies and programs, such as the *Forest Practices Code*, although these would ultimately become essential components of the provincial land-use strategy. This action-on-all-fronts approach underscored the need for coordination.

Commission workplan

Provincial direction

To establish strategic direction, CORE proceeded to

- Develop a *Land Use Charter*, describing principles of sustainability, public participation, and protection of Aboriginal rights as the foundations of the land-use strategy;

- Develop integrated land-use goals and strategic policies, in cooperation with government ministries, to give clout to the *Land Use Charter*; and

- Review the planning delivery system, with a view to improving its overall effectiveness and efficiency.

Regional plans

When the government established CORE, it also identified the first regions for the regional planning processes to take place. These were Vancouver Island, Cariboo–Chilcotin, and the Kootenays, the regions that had experienced the highest concentration and intensity of land-use conflict in the recent past. CORE initiated land-use planning processes in all three of these regions (the Kootenays region was subsequently split into two), with target dates for completion within 18 months.

Community-based participatory processes

To build on existing community initiatives and to test negotiated approaches to public participation in government decision-making at the local level, CORE initiated pilot projects in various communities around the province that had both a resource-management conflict and a willingness among all affected interests to participate. In addition, the commission cosponsored a workshop on local participatory processes for representatives of local round tables from around the province, to seek additional information and advice.

Dispute resolution

To design a dispute-resolution system, CORE reviewed existing appeal mechanisms and identified gaps in the current system. It also developed preventative approaches, particularly negotiations and mediation.

Aboriginal rights

To ensure that Aboriginal rights and treaty negotiations would be respected and not prejudiced, CORE developed, in consultation with First Nations, several discussion papers to identify options for First Nations' participation in developing the land-use strategy. In addition, CORE worked with government ministries to clarify the government policy that Aboriginal participation in the land-use strategy would not prejudice the treaty process.

Public participation

To ensure that the full range of affected interests and the general public would have meaningful opportunities to participate in the development of the provincial land-use strategy, CORE initiated a wide range of public participation processes. Its provincial-policy work was guided by multisector policy forums, open houses in communities around the province, and a variety of public discussion papers. To develop land-use and resource-management plans at the regional and local levels, CORE established a public negotiation process — called shared decision-making — on the basis of previous experience in the province, as well as that in the United States.

Independent oversight

Having established the provincial goals, the policy-development processes, and the regional planning processes, CORE identified a broad range of economic, social, and environmental indicators to serve as a starting point for developing a comprehensive sustainability-monitoring program.

Progress toward the provincial land-use strategy

By March 1995, CORE had completed six shared decision-making processes and numerous other consultative processes and published 17 public reports, including recommendations that covered all aspects of its mandate. In some cases, the shared decision-making processes generated full consensus on all issues. In others, the processes revealed substantial common ground and effectively identified the nature and scope of the government decisions required. Together with related government policy initiatives, the responses and decisions arising from the CORE processes and reports have resulted in substantial progress toward the development and implementation of the provincial land-use strategy. CORE's key policy developments, already made or pending, include those described in the following subsections.

Provincial direction

To consolidate strategic direction, the provincial government has taken the following steps:

+ Adopted the *Land Use Charter* as provincial policy;

+ Established the Land Use Coordination Office to coordinate the work of resource ministries and the implementation of regional and sub-regional plans;

+ Published a draft of the *Growth Strategies Act* to improve urban-growth management, including many of the proposed land-use goals and the mechanisms to improve the coordination of Crown-land and settlement-land planning;

+ Enacted a *Forest Practices Code* to improve and provide a legislative basis for forest practices;

+ Drafted the Forest Renewal Plan to dedicate a portion of forest-resource rents for reinvestment in provincial forests and forest-based communities;

+ Enacted the *Environmental Assessment Act* to provide a consistent cross-sectoral approach to assessing the impacts of site-specific projects; and

+ Established the Treaty Commission to oversee the treaty negotiations between First Nations and provincial and federal governments.

CORE has recommended an overarching *Sustainability Act* to provide a legal basis for the land-use strategy; to provide a secure commitment to sustainability; fair, open, and participatory decision-making; and enhanced coordination of government resource planning and policy development; and to ensure continual refinement and evolution of the strategy.

Regional plans

CORE completed the last of the four regional plans in October 1994. By March 1995, the provincial government had made its final decisions to implement plans in all these regions, based on the CORE recommendations. In total, these plans covered 20 million ha, almost 25% of the province. They incorporated 942 000 ha of new protected area, divided into 42 areas; 3.655 million ha of special-management lands; and 4.135 million ha of intensive-management lands. The remainder was dedicated to integrated resource management and settlement. (Special-management lands are areas where resource-management practices emphasize nonextractive values. Intensive-management lands are areas where increased resource production is to be emphasized, through enhancement activities.)

In less controversial regions of the province, an interagency subregional planning process is addressing land-allocation issues. About 60% of the province is now either covered under an approved allocation plan or in the process of plan development through CORE and interagency processes. The provincial government has also addrssed several long-standing issues in unpopulated parts of the province by establishing protected areas in the Tatshenshini–Alsek watersheds (an expansive wilderness area of 958 000 ha), the Khutzeymateen watershed (prime coastal grizzly bear habitat of 44 902 ha), and the Kitlope watershed (largest pristine temperate rainforest watershed in the world at 317 291 ha).

Community-based participatory processes

Drawing on its experience with pilot projects and extensive consultations, CORE recommended that the province encourage the regions to establish multisector, volunteer, consensus-seeking community resource boards (CRBs) to act as advisory bodies to government. CRBs are being established in all of the CORE regions as part of the implementation of the regional plans.

Dispute resolution

CORE is developing a mediation service to assist with dispute resolution and has recommended that the government rationalize and consolidate existing and required appeal functions and create a sustainability appeal board. Government has taken a initial step toward this by establishing a common secretariat for a number of existing appeal boards.

Aboriginal rights

CORE has recommended a Framework for Aboriginal Participation in land-use planning processes, designed to reinforce Aboriginal participation in planning while addressing interim issues and the transition to treaties.

Public participation

CORE has recommended that a general right of public participation be established in the *Sustainability Act* and that a code of conduct be adopted to reinforce effective and balanced civil communication in all participatory processes, as well as accountability and good faith in negotiations.

Independent oversight

CORE is increasing its emphasis on monitoring and oversight through the consolidation of a range of environmental- and resource-monitoring programs in its annual state-of-sustainability reporting.

Essential components of the provincial land-use strategy

Many of the essential components of the land-use strategy are already in place. The conceptual framework is now clear and consists of the following:

+ Provincial direction provided by a legal framework that comprises the principles, goals, and strategic policies needed to guide planning and decision-making for sustainability;

+ Coordination within and among the levels of government (federal, provincial, local, and Aboriginal) to ensure that the principles, goals, and policies for the management of lands and resources are effective;

+ A dispute-resolution system that ensures simple, accessible, and consistent review and appeal mechanisms and emphasizes the prevention of disputes through effective public participation and interagency coordination;

+ Participatory processes that provide meaningful opportunities for the public to help shape and inform decisions, as well as for building agreement among competing interests; and

+ Independent oversight that monitors the overall effectiveness of the strategy and reinforces continual adaptation and development.

Taken together, these components constitute essential tools for managing for sustainability.

Analysis

Shared decision-making

Meaningful public participation is an essential component of good representative government. In a complex world, in the midst of growing environmental, economic, and social pressures, people want to be closer to the decision-making process and have a say in defining the public interest, particularly where decisions affect them directly. Also, government requires informed public support to make the choices needed to move along the path to sustainability.

CORE's use of highly participatory, shared decision-making in its land-use planning is a step in the process of consultation, mediation, and evaluation that leads to a recommended regional land-use plan. This approach is attracting national and international attention because of the importance of sustainability and the universal need to find new ways to reduce land-use conflict.

Key concepts

Shared decision-making

Meaning

Shared decision-making is a consensus-based approach that jointly empowers those with authority to make a public decision and those who are affected by it to seek an outcome that accommodates (rather than compromises) the interests of all concerned. In a broader sense, the concept of shared decision-making provides for direct and effective public participation in government decision-making. It includes a spectrum of activities, from public consultations and reporting to open houses and town-hall meetings. It involves no formal change in governments' responsibility or legal authority to make decisions. Nor does it mean that all interests must reach consensus for the process to be successful. Rather, shared decision-making, by involving all those who are most interested in, knowledgeable about, and affected by the outcome, provides the opportunity for the final decision to be as well informed, balanced, and stable as possible.

Important elements

The cornerstone of a shared decision-making process is its cooperative problem-solving orientation. All parties' motives for cooperation stem from their realization that their goals are interdependent. Invariably, one party cannot get what it wants without the support or action of the others. Underlying the process is the assumption that by working with the other parties to solve a jointly defined problem, each party will gain more than it could by relying on other methods of influencing public policy (political bargaining, lobbying, campaigning, boycotting, litigating, etc.).

A shared decision-making process presumes that the decision-maker (usually a public agency) will work on an equal footing with representatives of those affected by the decision to negotiate outcomes acceptable to as many as possible. The opportunity to make decisions on specific issues shifts temporarily to the representative group, which includes the decision-maker. When consensus is reached, the expectation is that the group's decision will be implemented.

A key to success in shared decision-making lies in structuring the process so that it involves the affected parties' representatives in the design and development of the process, as well as in the negotiations on substantive issues. The participants must be involved from the stage of initially assessing whether it is appropriate to use a shared decision-making approach, right through to the ultimate implementation of the agreement and monitoring of the outcome.

The tool used to reach agreement in a shared decision-making process is interest negotiation.

Interest negotiation

Negotiation is a process in which the parties to a dispute design their own solution. The process may take the form of competitive, positional bargaining (which focuses on winning and losing and often results in either deadlock or compromise), or it may involve a cooperative problem-solving approach referred to as interest negotiation (also known as interest-based negotiation, consensus building, collaborative problem solving, cooperative negotiation, and principled negotiation). An interest-negotiation process is a structured, deliberate attempt by the parties to a dispute to cooperatively seek an outcome that accommodates the interests of all concerned.

The parties in a negotiation are usually quick to state their ideal outcomes. These ideal outcomes are the negotiators' respective positions. Negotiating over positions requires compromises that no side wants to make, particularly when important social, environmental, and economic values are at stake. Interest negotiation offers an alternative.

Interests are the motivating goals and objectives (that is, the needs, desires, concerns, and fears) that motivate the positions negotiators take. Interest negotiation encourages the parties to set aside their bargaining positions in favour of their underlying interests and work together to package those interests in an agreement that best satisfies the concerns of all.

Process framework

A complex, multiparty, shared decision-making process generally includes the phases described in the following subsections.

Before convening the participants

Preparation
The purpose of the preparation phase is to

+ Lay the groundwork for subsequent phases of negotiations by
  ⋄ Clearly establishing the objectives, scope, and methodology of the process,
  ⋄ Clarifying the guiding policy framework, information, and technical support needed for decision-making and determining whether they are in place,
  ⋄ Ensuring that the necessary incentives and methods for agency cooperation and the supporting financial resources are in place,

❖ Beginning the process of building constituencies and establishing communication links,

❖ Establishing personal, organizational, and procedural credibility with the prospective participants, and

❖ Communicating the nature and extent of the process to the general public; and

✦ Establish a basis for focused public participation by clearly defining which elements of the process are a given and which ones are negotiable and what information and policy direction the parties have to work with.

Assessment

The purpose of the assessment phase is to

✦ Assess (using readiness criteria) the preparedness of the decision-maker and the affected parties to support and participate in the round-table process;

✦ Enable prospective participants to

❖ Assess in a fully informed way the appropriateness and feasibility of using the interest-negotiation approach (after the participants convene, they will be encouraged to jointly assess the appropriateness and feasibility of this approach and formally agree to proceed),

❖ Fully canvas their alternatives to a negotiated outcome,

❖ Explore the willingness of their constituents to support the round-table process, and

❖ Help identify those who must be involved;

✦ Permit the decision-maker to assess the appropriateness and feasibility of a negotiated approach; and

✦ Permit an objective third-party assessment of the appropriateness and feasibility of a negotiated approach.

After convening the participants

Process design

The purpose of the process-design phase is to

✦ Create a suitable forum for shared decision-making and a workable process to support it (including rules to govern the process and carefully structured communication links between representatives and their constituents);

✦ Engender understanding and allow working relationships to evolve (help nurture a climate for cooperative negotiation);

+ Enable the participants to select a mediator to facilitate the process; and

+ Enable the participants (in consultation with the decision-maker) to reach a clear understanding of the conditions that will govern the implementation of a negotiated outcome.

Agreement building

The purpose of the agreement-building phase is to

+ Give the participants the opportunity to reach an agreement on a package of recommendations that accommodates (rather than compromising) the interests of all concerned; and

+ Secure broad-based constituency support for the agreement reached.

Negotiating the substantive issues in a shared decision-making process requires the participants to

+ Identify clearly the issues they wish to resolve (within the terms of reference);

+ Convert their positions on the issues into fundamental needs and interests;

+ Create a variety of options capable of satisfying the interests of all;

+ Craft an agreement based on objective criteria;

+ Assess whether the agreement can be implemented;

+ Formalize the agreement; and

+ Seek ratification from their constituencies.

After an agreement is reached

Implementation and monitoring

At the final stage, means are established for implementing and monitoring the negotiated agreement. Implementation and monitoring requirements are often part of the overall agreement on substantive issues. Land-use agreements typically require further negotiation during this phase. Effective implementation and monitoring provisions should, therefore, include procedures for any further negotiations to ensure the enduring success of the agreement.

The objectives of this final stage are to

+ Identify and agree on implementation requirements;

+ Establish a means for monitoring the implementation of the agreement and revising it when required;

+ Ensure broad participation in implementation, thus promoting shared responsibility and accountability; and

+ Establish mechanisms to identify and address newly emerging issues.

Implementation and monitoring are generally the responsibility of the decision-making authority. However, the negotiation group may be directly involved through periodic meetings to review progress.

Limitations

Not every shared decision-making process will result in full consensus. Much depends on external factors, such as a guiding government policy framework, necessary information, and sufficient time and resources to support the process. However, policy, information, and resources alone are insufficient to get the job done. At least as much depends on the participants' desire to work together to create opportunities for innovative outcomes that are not possible for any participant acting alone. Progress will be slow if some appear to themselves, or to others, to be able to get what they want independently.

Even if full agreement cannot be reached, the participants can still help by clearly defining problems, narrowing the scope of issues, and identifying a range of possible alternatives. Building working relationships and mutual understanding will in any case help ensure better outcomes.

Interest negotiation and learning[1]

The participants in interest negotiation require specific orientation in this negotiation process to communicate and work well with each other. In turn, substantive learning occurs through collaboration and through iteration of the newly learned process. They learn the necessary skills by analyzing their own self-interests; articulating this to others with different interests, perspectives, and needs; listening to others articulate their self-interests; jointly identifying fundamental principles, developing new analytical tools for measuring the advance of their own interests and the overall impacts of alternative solutions; defining a common problem; and collaboratively designing win–win solutions. Identifying such solutions is an iterative process of proposing, debating, analyzing impacts, and comparing alternatives.

Improving governance through shared decision-making

Shared decision-making gives government the opportunity to draw on as broad as possible a range of perspectives, interests, and experiences as it develops and implements public policy. This can supplement traditional representative

---

[1] The analysis in this section draws on the learning theory described by Bernard and Armstrong (this volume).

government by closing the distance between those making decisions and those most affected by them and by directly addressing the feeling of alienation that is a major dysfunctional force in conventional decision-making processes.

The consensus reached through public-policy interest negotiation, with the broadest range of interests effectively represented in a balanced process, is politically irresistible to government, and involves no formal devolution of decision-making authority. Even when an interest negotiation fails to achieve consensus on every issue, the participants and decision-makers are nevertheless exposed to a broader range of interests, better information, comparative analysis, and substantive learning. This inevitably results in more balanced, better informed, and more stable decisions, as they involve the people most interested in, knowledgeable about, and affected by the outcomes.

To make interest negotiation most effective in public-policy development, government should be directly involved in shared decision-making on a specific set of issues for a specific length of time and join with those affected by the decisions to reach consensus or, at least, to make a better informed decision. Because this process helps reconcile often inherently competing objectives, it is a particularly powerful tool for integrating social, environmental, and economic interests. Shared decision-making requires top-down statements of principles or goals, together with bottom-up, highly participatory public negotiations. Achieving social, environmental, and economic sustainability requires a continuing, dynamic relationship among the full spectrum of interests as they test policy decisions throughout their implementation against new information and experience. Relationships among competing interests developed through the shared decision-making process promote stability, as the balanced collaborative process and the relationships increase understanding and respect and promote flexible adjustments to new information and experience. A sustained process thus contributes to a sustainable result.

The learning process in shared decision-making tends to be irreversible, as the participants develop understanding and respect for other perspectives, as well as familiarity with a powerful tool for understanding their own interests and the best ways to achieve them through cooperation. The CORE experience in British Columbia, despite high levels of continuing conflict and dissatisfaction, shows that participants in large-scale, regional, shared decision-making processes continue to speak in interest-based terms and demand that the processes be continued at local levels for implementation of the regional plans, with local-level planning and continuing citizen oversight.

The shared decision-making approach provides a powerful motivation to learn about one's self-interest through careful analysis of alternative processes (litigation, civil disobedience, lobbying, media campaigns) and to learn about the interests of those who can either advance or block one's own interests. This not only leads to substantive win–win solutions but can also promote strong relationships among previously competing groups.

The shared decision-making approach to social, economic, and environmental integration can vastly improve governance by minimizing the dysfunctional distance between decision-makers and the people affected by their decisions and by addressing the often unrepresentative nature of governing minorities.

Readiness

A collaborative, learning-based approach to public-policy development requires a certain level of readiness on the part of all participants, including government. The mandate of the process should be clearly defined and constrained by a policy and fiscal framework that directs parties to a realistic and acceptable solution. Such a process also requires information; technical and administrative support; and a thorough prenegotiation assessment to ensure the involvement and cooperation of all the interested parties.

But it is important not to be too rigid. The learning in shared decision-making means that both substantive policy and procedural rules can be improved through the interaction of the participants. For example, the participants in the Vancouver Island regional negotiation process were unable to reach consensus, largely because government supplied no economic-transition policy. However, this lack caused the negotiators to develop and deliver to government a detailed, balanced, and highly practical transition strategy that was later adopted by government in the implementation of the final plan. Also, the parties' mistrust of each other and of those in authority meant that the process and procedural rules simply had to be negotiated and improved through trial and error across a range of negotiating processes. This created a body of experience in the province that is now a standard model for any new negotiations.

Often, to ensure that all interested parties see a negotiated approach as their best alternative, government has to make it clear that change will occur with or without one's participation. Also, as different groups are unlikely to be at the same stage of readiness and have the same ability to participate effectively, a convening or facilitating body, such as CORE, must ensure balanced participation. This can present a difficulty for the facilitator if its efforts are interpreted as being unfair by parties losing a previous power advantage.

Another challenge pertaining to readiness of participants can arise if they so mistrust each other that they establish process rules that effectively hamstring their ability to reach a consensus. This happened in the CORE Cariboo–Chilcotin region, where the decision-making rules required full consensus on the whole package, covering every issue. The participants reached agreements on many issues that they later had to abandon because of their lack of agreement on a total package. Although the facilitator might have imposed a more flexible set of process rules, the participants' mistrust would likely still have led to the collapse of the process in this case. A situation like this simply has to be allowed to run its course and ultimately be viewed as a learning experience.

The tension between readiness, achieved only through top-down imposition of substantive and process rules, and the messiness of learning as you go can also be a challenge to the political resolve of government in continuing to support shared decision-making processes through a period of learning and change. However, despite high levels of conflict and dissatisfaction with the failure to reach consensus in the initial processes, the public shows no inclination to abdicate the right to participate directly in major public-policy decisions — in fact, just the contrary.

Readiness is an issue at the outset of a negotiation process and throughout it. Learning takes time, and the time required varies among participants. In British Columbia, the completion of regional land-use plans seemed so urgent that government placed arbitrary time limits on the negotiations — an average of about 18 months. But these deadlines may have constrained the learning process and the participants' ability to reach consensus.

Compromise

Shared decision-making does not mean that parties must compromise or give up their fundamental interests and values but that they distinguish their fundamental interests and values from those that fail to take account of the needs of others. As different groups value things differently, self-analysis, communication, and understanding of each other's interests are needed to arrive at a mutually beneficial solution.

Another example from British Columbia concerns the issue of timber harvesting. Broadly stated, the interests of environmentalists are to protect biodiversity and to ensure that other nontimber values, such as visual quality, recreational use, and aesthetics, are sustained. Labour interests want secure employment; rural communities want economic diversification and stability; and corporate interests want the certainty of return on investment. Shared decision-making could produce a win–win solution: more sensitive harvesting techniques could lead to smaller-scale operations and employment through reforestation projects; a more diversified rural economy could be achieved through protection of fish stocks and tourism values; and corporate certainty could be achieved through reduced conflict and more-sustainable fibre flows.

Conflict can be debilitating, but it can also be a creative force if it induces an enriching variety of perspectives, broader understanding, and balanced and synergistic results through collaboration. Crisis can lead all interests to accept a new approach, but it can also challenge them to carefully examine their own underlying assumptions, values, and beliefs in the light of better information and new ideas and see whether adjustments are needed. For example, in the forestry dispute, participants recalculated the long-term sustainability of current harvest levels and compared the policy of total protection of the wilderness with the alternative of developing a new *Forest Practices Code* with less protection but with effective management for all values across the whole landscape.

With shared decision-making, parties in a public dispute do not have to abandon their key values, but the crisis should cause all parties to reconsider and perhaps temper their bedrock principles in light of their new understanding of other perspectives and needs and the clear inadequacy of the status quo that gave rise to the conflict in the first place. It is healthy to regularly challenge our first principles — comparing them with other legitimate points of view, new information, and continuing experience — to either renew our faith or make realistic adjustments. To do so effectively involves an intense learning experience.

Fundamental to the prospect of learning in shared decision-making processes are respect for the other parties and bargaining in good faith. CORE developed a code of conduct to guide these processes to this end, but a key role for the mediator or facilitator is to oversee this and to ensure balanced and effective participation. Effective collaboration occurs only if parties are willing to learn from each other, and learning only takes place if the parties are willing to understand and articulate their own interests honestly and to listen to others and understand their points of view.

Process integrity

Shared decision-making requires that at the outset, all parties, including government and nongovernmental participants, clearly understand each other's roles and responsibilities. Participants must recognize that their role is to advise the government decision-maker, but government should also be prepared to implement a consensus decision to the greatest possible extent. Government must be represented corporately as one of the parties to the negotiations and forthrightly inform the other participants of policy and fiscal constraints. The default procedure if no consensus is reached must be clearly understood from the outset — in the British Columbia case, CORE prepared and published a public report. A major threat to such processes will occur if government is insincere in its desire to involve and listen to a full range of public interests in the development of public policy. Objectives, expectations, and a code of conduct should be made explicit at the outset, and the whole shared decision-making process should be transparent to ensure a positive outcome, with or without consensus.

Such transparency will ensure a congruence between the espoused theory of public participation and how it actually plays itself out in practice. To encourage good-faith bargaining by all, government must not allow parties to make "end runs" around the process by directly lobbying ministers. The learning required to support a successful shared decision-making process requires at least a wary trust. This may be particularly difficult when the public's demand for greater participation is rooted in strong feelings of alienation, cynicism, and conflict. However, where expectations are raised and not met, government should expect a major backlash from people who feel betrayed.

Public negotiations and margins for learning

As Bernard and Armstrong remark (this volume), learning requires a margin for testing beliefs and assumptions against new information and ideas. This can involve risk, and although it seems that people in desperate situations can ill afford not to take a risk, a learning process is probably more successful if a margin is effectively built in to allow people to feel protected from threats to their fundamental interests.

In CORE public negotiations, the need for a government policy framework to provide protective margins was evident. For example, it was very difficult for labour leaders to accept new protected areas that could lead to a decrease in the timber harvest and therefore in the number of jobs, without an economic-transition strategy to increase employment intensity and opportunity through other means. Similarly, it was difficult for environmentalists to accept fewer areas of total protection without a strict *Forest Practices Code* showing how the full range of nontimber values could be protected through enforcement of higher harvesting standards. Forest companies were unwillingly to accept reduced harvesting areas without knowing that a compensation policy would apply to lost harvesting rights and that the long-term certainty of permanently designated harvesting areas would be assured. The lack of clear policy in all these areas made the first round of public negotiations difficult, but the processes themselves gave government clear direction on the issues and details to be included in provincial policy. Again, the top-down–bottom-up interaction contributed to effective public-policy change, development, and implementation.

Pilot projects are effective ways to test whether the risks of participation in public negotiations are manageable and whether the government is negotiating in good faith and has the competence to implement an agreement. Given the current levels of public cynicism toward government, simple statements of intention or political promises will not be believed: government has to demonstrate effective implementation and sustained commitment if the public is to be convinced that the risks are manageable.

The risks can also be better managed if the parties do not abandon alternative methods for getting their own way, such as government lobbying, media campaigns, threatened market boycotts, and lawsuits. Although good-faith bargaining may require the suspension of these activities, parties will probably, and appropriately, keep their options open as a defence against the risks of the public-negotiation process. Moreover, as the parties realistically assess their options they often discover that significantly higher risks accompany their alternatives to negotiation. CORE's experience has been that the parties often overestimate what they can realistically achieve by forcing an issue. As well, mediators can expand the margin for learning by encouraging the parties to engage in sober second thought about their alternatives. Also related to risk is the imbalance in resources and therefore negotiating power that parties may feel in entering into public negotiations. The facilitator must ensure that the power is effectively balanced through procedural rules, sharing of information, and financial assistance for participants.

Government officials also encounter risks in entering into shared decision-making negotiations, as they may fear that their statutory authority is being reduced. This can be overcome by demonstrating the empowering nature of consensus decision-making in delivering to statutory decision-makers creative, integrated, and widely supported solutions, in contrast to what they can expect from the weaker command-and-control decision-making processes. Professional development, interest-negotiation orientation, and cultural change within government bureaucracies are needed to reduce this feeling of risk.

In a shared decision-making process, procedural safeguards, such as straw polls, package deals, and access to more accurate data and information, can reduce risk. At the end of the day, incrementalism in building agreement over time can be much more effective than an all-or-nothing agreement (like the one required in the CORE Cariboo–Chilcotin negotiations, for example). Risk can also be reduced by neutral methods for obtaining information and clearly understood and supported analytical tools to measure the impacts of proposed scenarios. In the CORE regional negotiations, a land-use designation system, multiple-accounts analysis, and sectoral interest statements and maps contributed to a greater feeling of confidence in the learning process. The shared decision-making process will identify areas of common interest or areas where interests are valued differently and can be traded within a package solution. This is particularly important when the various interests are all legitimate, compelling, and apparently competing but really interdependent. Because they are interdependent, no party can achieve its interests without the support of the others. The collaborative approach of interest negotiation provides the opportunity to balance apparently competing priorities and arrive at a package solution.

If a clear crisis emerges, the status quo cannot be maintained, and change must be managed, the incentive to take a risk may bring all parties to the negotiating table. But government must make it clear that collaborative negotiations are where the action is and that failure to take part in these carries the greater risk of having no opportunity to influence the outcome. This is a fundamentally important factor in the success of the CORE negotiations. In particular, the motivation for the politically and economically powerful to negotiate with those without power has to come from a clear government message that the rules are changing and from a realization that they need the support of the others to get what they want.

Win–win solutions will be possible only through a process that allows the parties to distinguish items that are nonnegotiable from ones that may be desirable but are less important.

Unpredictability

Developing an integrated and sustainable social, economic, and environmental policy requires public participation. Sustainability cannot be imposed from above. It must be informed in a balanced and ongoing way by the widest spectrum of interests and perspectives. Moreover, the uncertainty and messiness of this dynamic process are also its major strengths, as it is likely to be more realistic and flexible if it responds to unforeseen circumstances and new information over time.

There is a superficial attractiveness to logical, sequential, strategic planning, commencing with top-down development of principles and goals, followed by establishment of a clear and comprehensive policy framework, detailed inventories, and technical support mechanisms, followed by broad-based and balanced public participation processes. But this is not the way the political world works, nor is it likely the way that learning most effectively takes place. CORE's experience with the development and implementation of integrated and sustainable social, economic, and environmental policy shows that the need for the policy is usually so urgent that all of those steps proceed at once. Although this approach can produce frustration, continuing conflict, and threats to political resolve, it also allows each aspect to inform and benefit from all the others, contributing to a more resilient, comprehensive, and often unexpected result.

Lessons from CORE

Public participation in policy development and implementation through interest negotiation promotes self-analysis, communication, and creative synergies, all of which stimulate learning and result in more balanced, stable, and wiser decisions, regardless of whether consensus is achieved. This shared decision-making process involves dynamic research: participants test their attitudes and ideas, first against other perspectives and in the context of increasing information, and second, against the experience of implementation. As Bernard and Armstrong (this volume) observe, new attitudes, values, relationships, and solutions develop particularly when learning takes place under the imperative of a changing and threatening environment.

CORE's experience to date indicates that even in the case of very considerable consensus, as in the East Kootenays, participants will eventually reach a point of impasse, and further agreement becomes impossible. When this happens, it is important to appreciate the progress that has been made and realize that the balance of interests and the new information, ideas, and options passed on to the statutory decision-maker will lead to better decisions. CORE's experience shows that up to the point of total impasse, however, a number of options are still open for moving the process beyond barriers. Such options include referring matters to technical working groups, involving third-party arbitration, asking parties to stand aside rather than impede consensus, developing a range of options, and voting (it should be understood that simple majority votes will be less persuasive and certainly not considered binding by the decision-maker).

The Vancouver Island and Cariboo–Chilcotin CORE processes came to a close because of the absence of a government economic-transition strategy to reduce the risk for rural communities and resource workers and because of the lack of definition of protected-area-strategy boundaries and percentages; the negotiations simply had to be concluded. But the missing policies were then recommended in the CORE report and later confirmed in the government decision.

A key aspect of interest negotiation is that each party must fully analyze its best alternative to a negotiated agreement (BATNA) and retain the right to withdraw from the process if it feels it can better meet its interests in other ways. But if these processes are going to work to produce considerable consensus and decisions that will be implemented, it is important for government to understand that it can affect various BATNAs through its policy development and rejection of lobbying initiatives outside the process.

Major lessons for all parties in CORE's regional processes were that the status quo was unsustainable; that change was occurring and was going to be managed through the development and implementation of an integrated and sustainable social, economic, and environmental policy within a comprehensive regional strategic plan; and that all interested parties were to be given an opportunity to take part in negotiating the plan. However, with or without consensus, at the end of the day CORE would be responsible for developing and publishing a recommended strategic plan and government would make the final decision on it. In retrospect, it is clear that some parties never believed that this would happen and thought that they could either maintain the status quo or influence a plan tilted to their advantage by withholding consensus or acting outside the planning process. Government actions in approving and giving legal status to the strategic plans has disproved these assumptions, and this should be a significant lesson to all interested parties in subsequent negotiations.

By every indicator of change — including percentages of different land-use designations, management objectives and guidelines, economic-transition and -diversification strategies, provincial-policy recommendations, and implementation strategies — CORE's recommendations and the subsequent government plans in all four regions have been a major departure from the status quo and in favour of sustainable resource and environmental management. Some land-use designations in the CORE recommendations were reconfigured in government plans, but they are almost identical in terms of sustainability principles, balance, strategic policies, and resource strategies. The similarity between the CORE recommendations and the government plans is best appreciated by comparing either one with the status quo, which demonstrates the extent to which CORE is introducing a new paradigm.

Meaningful public participation, introduced through the CORE processes, was also adopted by the CRBs, which were developed by the regions to
oversee the implementation of the regional plans, advise government on local
resource planning, participate in economic-transition initiatives, provide a community link to Aboriginal treaty negotiations, and advise government agencies
on a range of social-, economic-, and environmental-sustainability issues.

Another important lesson relates to the role of professionals and technical
advisors in these CORE processes. Each negotiating table had a team of government technical advisors, assigned from a range of government agencies. This
approach had the very positive impact of requiring government experts to work
as an integrated team, mostly for the first time. In doing so, they had to subor-
dinate their individual ministry mandates to support the objective of the process
and develop collaborative and balanced solutions. In addition, their demonstrated competence and diligence gained the respect of the wide range of sector
participants, enhancing the credibility of government in the eyes of the public.
In turn, these government officials gained a respect for the volunteer efforts and
wisdom of the different perspectives brought to the processes by members of the
public and came to appreciate the value of public participation as an empowering force for government, rather than a threat to their authority.

Implementation of sustainability

Achieving sustainability through the implementation of integrated social, economic and environmental policy is a highly dynamic undertaking. CORE's
experience shows that it is a continuous process of public involvement in advice
and oversight at the local level; development of sustainability indicators; monitoring and enforcement of standards; public review and reporting; and ongoing
amendment of policies and plans to respond to new experience. This is an adap-
tive management approach, with permits and approvals issued to resource users
on a performance basis.

Sustainability is a product as well as a process. It is based on balancing
social, economic, and environmental principles and integrating goals. These
often compete and therefore must be reconciled through highly participatory
planning and subjected to a continuous, dynamic process of measurement and
adjustment. Broad interest negotiation and public participation lead to better-
informed, balanced, and stable decisions and, because of the understanding and
respect developed among the parties, provide the government with the flexibility to respond to new information and experience.

CORE's experience shows that the gap found sometimes between the
development of policies and plans and their implementation can help to mobilize constituency support from a wide range of interest sectors for an emerging
consensus, with a view to giving the plan time to demonstrate its effectiveness.
However, great care must be taken to ensure that the disparity between policy

and practice is not promoted in bad faith to discredit the process and the result. The fact is that such disparities are inevitable in sustainability planning at a time of rapid and threatening change and many unknowns.

Incentives from government to motivate participants to adopt sustainability plans have proven not only helpful but essential in the CORE processes. The Forest Renewal Plan, for example, drawing on the large financial resources made available from increases in the rent on public land that is used for timber harvesting, is providing for forest rehabilitation, increased and diversified employment, value-added manufacturing, intensive forest management, and research and development. These, in turn, provide the motivation for a wide range of interests to support the transition to sustainability. Enlightened self-interest can be consistent with good public policy.

The success of implementation of an integrated sustainability plan must be measured against the integrated principles and goals, rather than by the letter of the agreement or plan. New experience and information may change the details but not the need for a balanced and participatory approach.

Communications policy and community involvement

A major challenge in the CORE regional planning processes has been to ensure meaningful, broad public participation in public-policy development. This meant helping to organize interest sectors that could be represented in an accountable way at the public negotiation sessions. A major challenge and often a failure of the initial CORE processes was to keep the sector constituencies and general community informed about the process and involved in guiding the results. It became very clear that greater effort and resources were needed to ensure effective constituency and community participation.

A major difficulty in the large-scale CORE regional planning processes was that many of the sectors were not natural constituencies and covered such large geographic areas that representatives experienced great difficulty keeping their constituencies involved. However, this challenge must be balanced against the value of breaking down the intransigence of organized interest groups by forming sectors with broader perspectives. A further issue was that although the negotiation sessions were held regularly throughout communities within each region, discussions of land-use planning were often too detailed and tedious to catch people's attention, even with public sessions being held in their very midst.

In response to this experience, CORE developed a code of conduct, with specific recommendations for sector representatives' accountability to and communication with their constituencies, as well as their responsibilities to communicate the whole negotiation process to the general community. Learning from the previously concluded Vancouver Island and Cariboo–Chilcotin processes, negotiators in the West and East Kootenay CORE processes took their table reports to open houses in communities throughout each of their regions before CORE prepared its final recommendations for government consideration.

Future strategic-planning processes will likely take place on a smaller geographic scale and use existing sector constituencies, to encourage better communication. The CRBs are expected to be standing volunteer bodies that will develop a close relationship with the general community on a range of issues.

Conclusion and application

Public participation developed through multisector interest negotiation is essential to managing for sustainability. Such negotiations not only enable government to obtain the kind of comprehensive and balanced information needed to develop and integrate economic, social, and environmental policy but also encourage the stability of the integrated policy, as it is perceived to be rooted in and to reflect the broad public interest. By encouraging conflicting interests to understand and reconcile their differences, the process also builds goodwill and resilience within communities. This is in stark contrast to consultative models that can exaggerate the differences among conflicting interests and cause participants to adopt extreme positions in the hope that a compromise decision will be in their favour.

The effectiveness of interest negotiation in this context is directly related to the extent of development of four other essential and interdependent components in any comprehensive land-use strategy:

+ Government direction in policy development;

+ Coordination of initiatives throughout the institutional structure of government;

+ A comprehensive system to resolve land-use and resource-management disputes; and

+ Continous, independent oversight of implemented sustainability initiatives.

The necessary haste with which economic, social, and environmental policy is being integrated underscores the need for careful analysis of the successes and difficulties experienced in that integration. In British Columbia, CORE conducted such analyses periodically while developing the land-use strategy and facilitating the interest negotiations. CORE has drawn on this analysis to prepare a conceptual workplan for integrating economic, social, and environmental policy through shared decision-making (Table 1).

Although there is no single, universally correct way to develop and deliver a shared decision-making process, in practice, mediators tend to work with the sponsoring organization and the prospective participants to design a process to suit specific circumstances. Complex public-policy negotiations generally require a series of interrelated agreements dealing initially with process and

Table 1. Workplan for integrating economic, social, and environmental policy.

Process steps	Product
Defining the policy context for the process	
Prepare reference material for prospective participants describing the work of the task force, including study methodology and expected products	Participant reference material
Develop procedures for economic-, social-, and environmental-impact assessments	Impact-assessment procedure
Review participant-assistance program and funding levels	Participant-assistance program
Work with the appropriate government agencies to prepare economic-, social-, and environmental-policy background papers to guide and support the round-table process	Policy base
Developing terms of reference	
Assemble and task process-management team, including process manager (convener or mediator) and secretariat staff (to coordinate technical, communications, and logistics or funding needs) — clarify roles and responsibilities	Process-management team
Assemble and task technical-support group, ensuring field capacity within supporting agencies — clarify roles and responsibilities	Technical-support group
Define a critical path for the study, including information assembly, approximate meeting schedule, decision milestones, and interim products	Timeline
Identify information requirements and timeline for baseline-information assembly — commence information gathering	Baseline-information assembly
Prepare provisional public-participation strategy, including draft terms of reference, range of involvement opportunities, prospective participants, decision structure, negotiation parameters, draft rules of procedure, and roles and responsibilities	Public-participation strategy
Develop program for public education and orientation	Planning and negotiation training modules
Develop communication strategy for public and local government	Provisional communication strategy
Identify and contact potentially affected parties — assess level of public participation required	Prospective-participant contact list
Develop process-readiness criteria, such as organized sectors, policy framework, process and procedure framework, well-defined study methodology and sequence, technical support, baseline information, sector negotiation training, and initial work on sector interests and goals	Readiness criteria
Prepare budget	Budget
Prepare information package for prospective participants covering terms of reference, policy background papers, baseline information, process framework, draft rules of procedure, draft workplan, decision structure, and roles and responsibilities	Comprehensive information package
Confirming and structuring the process	
Convene community information meetings to describe the initiative, policy framework, and range of opportunities for participation	Informed communities
Meet with affected parties to describe tasks, discuss expectations, acknowledge regional issues and concerns, and explain planning and process constraints	Informed interest groups

(continued)

Table 1 concluded.

Process steps	Product
Structure round-table membership	Sector representation
Adapt participation strategy to fit circumstances and complete process workplan	Participation plan
Deliver negotiation orientation to participants	Oriented participants
Work with sectors to assess participation and initially define sector interests and goals for the process	Sector commitment to proceed
Address conditional commitments from sectors (task-force assessment)	Task-force commitment to proceed
Convening the process	
Commence process with a meeting, perhaps structured as a public forum, to	Issue identification
✦ Allow representatives to identify and generally discuss key issues	Defined planning context
✦ Discuss and clarify study, information base, policy framework, and workplan	Mobilized public interest
Undertake joint assessment	Commitment to proceed
Discuss and adopt rules of procedure	Procedure agreement
Discuss and adopt communication strategy	Communication plan
Advise sectors on process administration and participant assistance	Participant-assistance agreements
Identifying and resolving issues	
Identify key issues, milestones, and interim deadlines (guided by terms of reference)	Negotiating agenda
Facilitate definition of sector interests and identify potential evaluation criteria	Definition of interest and decision criteria
Identify specific interest-related information requirements and timeline for assembly — commence information gathering	Interest-related information
Conduct joint exploration of interests and development of evaluation criteria	Comprehensive decision criteria
Develop specific action plans and strategies aimed at innovative, practical solutions to the key issues through a process of iterative development, impact assessment, and evaluation	Specific actions and strategies
Host public and sector information forums	Community dialogue
Agree on recommendations	Draft report
Arrange public and agency consultation, as required	Community and agency dialogue
Submit report	Final report
Implementing	
Monitor and enforce implementation	Enforcement
Support ongoing community oversight	Oversight
Develop indicators and independent reporting on progress	Progress indicators
Develop mechanisms for review, appeal, and amendment	Review and appeal mechanisms

procedure, information assembly, and direction (vision and goals) and then with the resolution of specific issues (defining sectoral interests and developing and evaluating options to accommodate the needs of the participants). This seldom constitutes a step-by-step progression. Negotiations more often follow an irregular path as the participants try to deal with the dynamics of conflict and the complexity of the task.

A public negotiation process requires leadership from the sponsoring organization to establish the overall direction and clear terms of reference for the process. Uncertain terms of reference confuse and frustrate participants and often sidetrack the process. On the other hand, terms of reference that are too constraining can limit the opportunity to accommodate legitimate interests and to generate innovative outcomes. It is important that prospective participants understand the nature and scope of the negotiations before they commit to participating in them.

Chapter 5

Planning Act Reforms and Initiatives in Ontario, Canada

G. Penfold

◇ ◆ ◇

Introduction

Historically, governments, businesses, and individuals involved in deliberate processes of development and change have focused on economic benefits and selected social benefits, with less regard for environmental and broad social impacts. Sustainability demands a new balance of the benefits and impacts of change, particularly as these relate to the natural and social environments. This chapter describes the reform of a land-use planning system, one of the key means of managing change, as an example of an attempt to better integrate the emerging values of sustainability with the traditional considerations of land-use planning. The resulting integration of social, environmental, and economic interests was part of both the reform process itself and the new policies and institutional arrangements that came out of that process.

A land-use planning system has been in place in the Province of Ontario, Canada, since 1946. The *Planning Act* established procedures and authority for making decisions about land-use change on private and municipal lands. Over time, several reviews and many amendments to the *Planning Act* resulted in a detailed and complex system of policies, procedures, roles, and authorities. In the last 20 years, this complexity led to increasing concern about the efficiency of the planning system and its effectiveness in addressing environmental and social impacts.

The idea of having a formal provincial policy to respond to these concerns began to be discussed in the 1970s, when the province established policies for the protection of agricultural land, mineral aggregates, and flood plains. These policies acted as guiding principles for municipal policies and related planning decisions.

Despite these policies, concern continued to increase about both the effectiveness and the efficiency of the planning system. In response, the provincial government appointed a Commission of Enquiry (the Commission on Planning and Development Reform in Ontario [CPDR]) in 1991 to review the planning system and to make recommendations for change. One significant outcome of this process was a *Comprehensive Set of Policy Statements*, approved by Cabinet in May of 1994. Another was a revised *Planning Act*, approved in December 1994. These changes came into effect in March 1995. A provincial election in May 1995 resulted in a new political party taking power. The new government reviewed and amended both the provincial policies and legislation in May 1996.

Case description

The planning context in Ontario

Ontario has a variety of planning and development contexts and political and administrative structures. At the beginning of CPDR's work, the non-Aboriginal provincial population of more than 10 million was served by 792 local municipalities. Seventy percent of these municipalities were in rural areas and had populations of fewer than 5 000. At the other end of the scale, the Metropolitan Toronto regional government served a population of about 2.4 million, and the City of Toronto had a population of more than 600 000.

More than 90% of the population lived in southern Ontario, in 12 regional- and 27 county-government structures. These municipalities had formed a second tier of municipal government, containing local municipalities. Almost 40% of the total provincial population lived in Metropolitan Toronto (22%) and the adjacent three regions (16%).

In northern Ontario, one-half of the 800 000 residents were living in six cities. The other one-half lived in small municipalities, with no second tier of municipal government. About 50 000 non-Aboriginal people came from unorganized areas, without a municipal structure. Some of these small municipalities and unorganized areas had appointed planning boards to deal with local planning matters; others had local service boards to provide basic services, such as roads and fire protection. In unorganized areas outside planning-board jurisdictions, the province administered planning.

History of planning in Ontario

The *Planning Act* of 1946 established the authority for municipalities or joint municipal planning areas to develop official plans and zoning regulations. The initial application of this legislation was in cities and surrounding areas. As rapid growth and development occurred through the 1960s, planning became established in most municipalities. Managing this complex system started to become an issue of public concern. A review of the planning system in 1971 by the Ontario Economic Council recommended, among other initiatives, "a basic policy on the allocation of provincial resources" and "a consistent philosophy on critical policy concerns including particularly environmental conservation, social and economic welfare, and community amenity" (OEC 1971, p. 97). This review did not result in changes that supported the idea of integrated policy.

A review of the *Planning Act* by the *Ontario Planning Act* Review Committee (PARC) in 1977 was more successful. PARC (1977, p. 30) recommended that legislation define provincial interests to include

> *the distribution of economic and social resources among the residents and regions of the Province; the maintenance of the province's agricultural and rural base; and the distribution of activities which have an "undesirable" local impact but are necessary from an overall Provincial standpoint.*

Low-income housing and gravel extraction were used as examples. PARC also "expected that the principles will be elaborated from time to time as specific provincial policies." These policies "should be implemented by way of regulations or other statutory orders, formally adopted by the provincial Cabinet and having the force of law."

Subsequently, in 1983, the province amended the *Planning Act* to include defined provincial interests and procedures for the review and approval of policies by provincial Cabinet. Under the 1983 *Planning Act*, every municipality and planning board "may" develop an official plan to "provide guidance for the physical development of the municipality" while "having regard to relevant social, economic and environmental matters" (Ontario 1989, p. 3). The Minister of Municipal Affairs is given the authority to approve these plans. These official plans and zoning bylaws control private development and guide the planning and development of municipal infrastructure.

Provincial policies are implemented through a requirement in the legislation that all decision-making authorities, including local governments, are to "have regard to" provincial policies in their decisions (Ontario 1989, p. 4). All official plans are to be approved by the Minister of Municipal Affairs. Amendments to plans and large-scale developments, such as plans for subdivision, also require provincial approval. If conflict occurs over the implementation of policies, appeals may be taken to the Ontario Municipal Board (OMB), which can generally make final decisions on planning matters.

Between 1983 and 1992, the province adopted four policies:

+ Mineral Aggregate Resource Policy (1986; Ministry of Natural Resources);

+ Flood Plain Planning Policy (1988; Ministry of Natural Resources);

+ Land Use Planning for Housing Policy (1989; Ministry of Housing); and

+ Wetlands Policy (1992; Ministry of Natural Resources).

These ministries, which all had related mandates, took responsibility for ensuring the implementation of policies through the review and approval of municipal plans and development decisions. Interest groups and the public were involved in policy development through a process of review and response to ministry policy proposals. This approach generally proved to be challenging and conflict-laden. A long period between the establishment of policy concepts and final approval was the norm. The Wetlands Policy, for example, took more than 10 years to move from concept to approval. Imposition of policy by the province could have substantially shortened this time frame, but such action carries with it the risk of negative public reaction and was therefore considered a politically unacceptable strategy.

By 1991, various ministries had adopted guidelines as an alternative to formal policy. These guidelines addressed issues, such as noise, distance separating industrial facilities from sensitive land uses, and protection of significant areas. Because guidelines could be approved without consultation, they avoided the problems of conflict and time delay. However, lack of public and political support for these guidelines resulted in conflict during the review of official plans and development applications, and the legitimacy of giving guidelines the force of policy was also questioned.

Within this general framework of policies and guidelines, review of plans and development applications was subjected to long delays. Review time frames of 3–5 years were common for major plans and development applications. An additional 2 years was required if an approval was appealed to OMB. The public continued to raise issues about protection of environmental and social well-being. As well, graft and corruption were suspected to be occurring at the municipal level; this resulted in an investigation being conducted by a special unit of the Ontario Provincial Police, with charges eventually laid.

To respond to these problems, the government appointed CPDR under the *Public Enquiries Act*. CPDR's challenge was to make recommendations for changes to legislation, policy, or both, to resolve these various difficulties.

The Commission on Planning and Development Reform in Ontario

CPDR was given a broad mandate. It was to recommend changes to the *Planning Act* and related policies that would restore confidence in the integrity of the planning process; better define roles, relationships, and responsibilities; and make the planning process more timely and efficient. A critical part of CPDR's mandate was to recommend changes that would better protect the public interest in planning and land development, including "environmental considerations." CPDR was directed to "consult widely, conduct research," "foster dialogue," and submit its final report by 1 July 1993 (CPDR 1993, p. 165).

The *Public Enquiries Act* gave CPDR the legal authority to access information and control public processes. CPDR also had more than adequate financial resources: its $2.8 million expenses were only about one-half its allocated budget. Its logistical and other support was provided by a special branch of the Ministry of the Attorney General.

CPDR's approach

CPDR's goal was to make recommendations that would generally be acceptable to the public and to those involved in planning and development and that would have a realistic possibility of being implemented. The task was to find common ground among the various stakeholders involved in planning. CPDR used a participatory and solution-building approach that involved planners, developers, citizen activists, environmentalists, farmers, municipal staff and politicians, provincial staff and politicians, and others who work with the planning process.

Between the start of its activities in September 1991 and submission of its final report in June 1993, CPDR organized 15 working groups to generate proposals for discussion. These proposals were published and circulated in newsletters. CPDR held 46 public forums on these proposals, through four rounds of formal public hearings across the province. Thirty-eight less-formal community meetings were also organized. CPDR attended more than 80 conferences and workshops. These were usually organized by stakeholders and interest groups. In total, CPDR talked directly with more than 23 000 people. It also met regularly with organizations and interest groups, including provincial ministries, agencies, and politicians.

CPDR had a mailing list with more than 19 000 names for circulation of newsletters and reports. An additional 5 000–10 000 copies of documents were distributed at conferences and meetings. About 2 100 written submissions were received, including 1 200 on a draft report released in December 1992. CPDR also deliberately used the media as a resource, including a biweekly interview on CBC's *Radio Noon* program in Toronto and Sudbury; press releases and news conferences at major steps in the process; and meetings with the editorial staff of the daily papers in the public-forum venues.

Results of the review

The final report of CPDR was released in June 1993. It contained 98 recommendations dealing with provincial policy, changes to legislation, and new administrative procedures and organizational arrangements. A key recommendation was that the province adopt a comprehensive set of policies addressing six policy areas. CPDR's recommendations also included implementation procedures and suggestions for resolving conflicts between policies (CPDR 1993).

On receiving the final report, the Ministry of Municipal Affairs reviewed and amended the recommended policies. The changes addressed concerns that the ministries and government felt were inadequately addressed in the recommendations. These changes were relatively minor. The Ministry of Municipal Affairs released a draft of the revised policies in December 1993 and asked for comment before the end of March 1994. About 600 written submissions on these proposals were received. After further changes, Cabinet approved a *Comprehensive Set of Policy Statements* in May 1994.

The approved policies covered the same six policy areas recommended by CPDR:

+ Natural-heritage, environmental-protection, and environmental-hazard policies;

+ Economic, community-development, and infrastructure policies;

+ Housing policies;

+ Agricultural-land policies;

+ Conservation policies; and

+ Mineral-aggregate, mineral-resource, and petroleum-resource policies.

A section on interpretation and implementation was also included. Legislative changes put in place a requirement that planning actions be consistent with these policies. The intention of this change was to strengthen the former requirement to "have regard" for policies. To ensure implementation, the province retained the authority to approve county and regional plans. These second-tier governments were given authority to approve local plans, which also had to be consistent with policies and upper-tier plans.

In May 1994, the legislature also gave first reading to Bill 163, which amended the *Planning Act* and four other acts related to planning. Between September and December 1994, a standing committee of the legislature reviewed Bill 163. This review included public hearings in 12 centres across Ontario. Bill 163 received third reading in December 1994. Proclamation of both the policies and the new legislation took place on 28 March 1995. Implementation guidelines were also released at that time. They included more than 700 pages of background, interpretation, and suggestions for implementation of the approved policies.

Subsequent changes

In the summer of 1995, after a provincial election, the Conservative party took power. It was elected with a mandate to control debt and deficit and to stimulate the economy. In this new political climate, social and environmental planning programs, policies, and regulations were seen as obstacles to economic growth. In early 1996, the recently adopted policies and legislation were reviewed through a process of circulating proposals and requesting submissions. A standing committee held hearings on new legislation, Bill 20: the *Land Use Planning and Protection Act.* By May 1996, revised legislation and policies were in place. Key changes included

+ Withdrawal of the implementation guidelines introduced by the previous government;

+ Removal of the requirement that the province approve upper-tier plans;

+ Reinstatement of the requirement that planning authorities have regard for provincial policies; and

+ New requirements for, and limitations on, public rights of review and appeal.

The new provincial-policy statement has three policy areas:

+ Efficient, cost-effective development and land-use patterns — developing strong communities, housing, and infrastructure;

+ Resources — agriculture, mineral resources, natural heritage, water quality and quantity, cultural heritage, and archeological resources; and

+ Public health and safety — natural and artificial hazards.

Most of the areas addressed in the previous policy statement were retained in this revised format. However, the philosophy and content were substantially changed to give municipalities much more discretion to interpret policies in a local and regional context. For example, the previous policy statement required an environmental-impact statement for development proposals that had potential impacts on environmentally sensitive areas. The new policies require proof that the natural features or ecological functions of the area will not be negatively affected. Similarly, the previous policy statement required that 30% of new dwelling units be affordable to households falling in the lowest 30th percentile of the household-income distribution in the area housing market. The new policies encourage housing forms and densities designed to be affordable to moderate- and lower-income households. A requirement to permit two households in each single-family dwelling unit was removed. The changes effected by the new government shifted the emphasis of policies and implementation away from environmental and social concerns and toward economic concerns and shifted implementation from provincial control to local control. But the comprehensive models for policies and the challenges of integrating economic, social, and environmental priorities in local and regional planning still remain.

Case analysis

Analytical framework

CPDR's objectives were both substantive (for example, protection of publicly valued goals concerning the environment and agriculture) and procedural (for example, improving efficiency, openness, and accountability).

Available literature suggests that the public and academic sectors carried out considerable research on most of these issues. A key finding from a review of this literature and the results of previous reviews of the planning system, in 1971 and 1977, was that the province needed to establish a policy as the basis for a strong planning system. Except for the four policies previously noted, the province had not developed an integrated set of policies.

CPDR's approach focused on developing recommendations with public, stakeholder, and political support. Changes in roles and procedures had to ensure that policy could and would be implemented. The aim was to use a consensus-building strategy to generate good policy recommendations and an agreement that policies should be adopted and implemented.

Integrative elements

The work of CPDR was limited to a review of the *Planning Act* and related policies and legislation. General social and economic policy fell outside its mandate. The land-use planning context, therefore, limited the scope of integration of social, environmental, and economic policies.

Within this context, however, social-, economic-, and environmental-policy components did emerge in several ways. In Bill 163 (Ontario 1994), a new section defined the purposes of planning. Two subsections are particularly relevant:

+ 4(1.1)a — "to promote sustainable economic development in a healthy natural environment"; and

+ 4(1.1)c — "to integrate matters of provincial interest in provincial and municipal planning decisions."

Section 5.2 of Bill 163 provided a definition of provincial interest. This definition covered such matters as protection of ecological systems — including natural areas, features, and functions — and the protection and conservation of agricultural and natural resources (environmental interests). It also covered the orderly development of safe, healthy communities and adequate provision of health, educational, social, cultural, and recreational facilities and a full range of housing (social interests), as well as protection of the economic well-being of the province and municipalities and the adequate provision of employment opportunities (economic interests).

The *Comprehensive Set of Policy Statements* elaborated on these interests. The natural-heritage, environmental-protection, and environmental-hazard policies require protection of water resources and natural features of significant interest, including wetlands, woodlots, and natural habitats. These policies also protect people from the consequences of development in hazardous areas, such as flood plains and areas subject to erosion or wave damage. The agricultural-land, mineral-aggregate, mineral-resource, and petroleum-resource policies address the goals of natural-resource protection. Conservation policies address issues in managing renewable resources, such as energy and water, and minimizing waste. Transportation components of this policy address both resource and social concerns. Housing policy addresses the need for affordable housing (social interests). Finally, the economic, community-development, and infrastructure policies address issues in services and infrastructure (linking social services and facilities planning to land use) and also support planning for a diversified economic base (social and economic interests). These policy areas are also included in the revised set of polices developed in conjunction with Bill 20.

Some of these policies are exclusionary (for example, no development in significant wetlands), and related land-use conflicts are relatively easy to resolve. However, other policies may conflict in specific situations, with no clear direction about which policy goal takes priority. For example, good-quality agricultural land can be located over good-quality aggregates. In this case, the policy-implementation process would have to define the highest priority in this context, or minimize conflicts through mitigation processes (for example, rehabilitation of the site to agricultural use after aggregate extraction), or both. Planning is primarily concerned with the identification and resolution of these conflicts. Thwarting the purpose of the planning process by ignoring or overriding one or another interest is much more difficult under a comprehensive policy framework.

CPDR's policy process

The political environment

Several factors in the external environment contributed to the success of CPDR's policy process:

+ The government in power (the New Democrats) was newly elected and was interested in change. When it was in opposition, it had been critical of the lack of attention to provincial interests in planning and supportive of better protection of the environment.

+ Several recent public reviews and appeals concerning planning problems had criticized the planning system. The Royal Commission on the Future of the Toronto Waterfront was one of these significant reviews. In one of its reports, "Planning for Sustainability," the Royal Commission stated that "a major weakness in the land-use planning system in Ontario is the provincial government's lack of leadership, coordination and direction in the land-use planning process" and that "the Province's first step in reestablishing leadership in this area would be to establish provincial interests by developing policies as envisaged and provided for under Section 3 of the *Planning Act*" (Doering et al. 1991, p. 81). The final report of the Royal Commission promoted the idea of an ecosystem approach to planning, with a particular emphasis on watershed planning (Crombie 1992).

+ The province was directly engaged in regional planning issues. The Provincial Office of the Greater Toronto Area had been working with

Toronto-area regions and municipalities to coordinate growth management and infrastructure policies. A similar exercise was under way to coordinate regional and local planning policies on the Oak Ridges Moraine, a significant natural area north of Toronto. The difficulty in providing coordination without a provincial policy framework was evident in both exercises.

✦ Ontario Ministry of Municipal Affairs and other ministries had been working internally on an umbrella policy, an initial attempt to create a comprehensive set of provincial policies.

Comparison with the traditional process

The traditional approach used by policymakers and by most commissions is to develop policy based on the perspective of the political and administrative systems of government, with limited public participation. The public is usually asked for opinions on issues and approaches but rarely has a chance to comment on policy proposals until the process is virtually complete. These processes often have open-ended time frames, with little public understanding of how or when decisions will be made.

The establishment of CPDR provided the opportunity for a different approach. The first challenge was to identify individuals to act as commissioners. The government asked John Sewell, former council member and mayor of Toronto and Chair of the Metro Housing Authority, to chair CPDR. Toby Vigod, environmental lawyer and executive director of the Canadian Environmental Law Association, and George Penfold, an associate professor at the University School of Rural Planning and Development at the University of Guelph, were also asked to be commissioners.

Before accepting the appointments, the three selected candidates met several times to establish a common understanding of the mandate and a general strategy for carrying it out. They asked for several changes to the mandate as a condition of acceptance of appointment: one of these was to add "the goals of land use planning" to the review (CPDR 1993, p. 165). This request resulted from an agreement among the candidates that policy would be key to any significant change in the planning system. This request also put the Minister of Municipal Affairs and Cabinet on notice that policy would be a focus of CPDR's work.

CPDR appointed Wendy Noble as its executive director. Noble was a manager in the Ministry of Municipal Affairs and had led the provincial umbrella-policy review. She had an excellent understanding of the interests and concerns of the various government ministries. Also, in appointing John Sewell to chair the commission, the government selected a well-known public activist, who was oriented to reform, a supporter of community interests, and a media figure. This appointment made it clear that this review was not another bureaucratic exercise and created an immediate public expectation of change.

Implementing the objective of policy and legislative change that would be generally agreed to among the various stakeholders — including the province — meant creating innovative public processes and forums to allow all interested parties to hear each other's concerns. Several approaches were used. First, CPDR established a specific schedule of activities. All stakeholders knew within 3 months of the start of CPDR's work that it intended to submit its final report by April 1993. A draft report was scheduled for December 1992. Three scheduled rounds of public forums would precede the draft report. This allowed participants to organize their resources to respond to CPDR's schedule.

Second, CPDR established working groups to generate ideas for discussion. The working groups comprised stakeholders dealing with the various planning contexts in the province: urban areas, urban-fringe areas, rural areas and small settlements, cottage country, and two groups in northern Ontario (east and west). The working groups typically represented the perspectives of ratepayers' associations, municipal administrators, municipal planners and politicians, provincial ministries, environmental interest groups, lawyers, First Nations, development interests, and economic-development planners. However, group members were asked to bring their individual views to the table, not those of an organization. CPDR selected group members based on their reputation for being skilled and thoughtful individuals respected by their peers.

The first six groups were asked to generate ideas about goals and policies for planning. These groups met in sessions of 2–3 hours each, for a total of about 12 hours. CPDR published and circulated the results of these deliberations in a newsletter, *The New Planning News*. Public forums in nine centres across the regions of the province followed in January 1992. People were asked to present either written or verbal comments on these proposals or their own ideas. CPDR consolidated the comments and prepared a comprehensive set of draft policies.

CPDR discussed these revised policies with the committees and task forces established by the working groups. After further revisions, the proposed policies were published in the April 1992 issue of the newsletter. Another round of forums followed. Revisions from this review and further consultation with interest groups resulted in revised policies, which were published in the draft report in December 1992. Comments from the final round of public forums and meetings with interest groups resulted in the policies included in the final report. These recommended policies were subsequently revised by the Ministry of Municipal Affairs and were circulated for comment in December 1993. The ministry provided a 3-month period for comments. After further revisions, a final set of policies was approved by Cabinet in May 1994.

Draft policies were also discussed on biweekly radio broadcasts, in news articles, at conferences and seminars, and at open public meetings. In sum, over a period of 2 years, the polices went through five stages of refinement and public review. A similar process was used to create recommendations on issues of planning process and development control.

Roles of beneficiaries, organizations, and institutions

Stakeholders played several roles. First, individuals with experience in dealing with stakeholder interests were members of working groups. In addition, most participating organizations formed planning-review committees or task forces. These committees met directly with CPDR to discuss their concerns, as well as presenting positions at the various public forums and interest-group seminars and conferences.

After the first round of working groups, CPDR formed a Leaders Group, consisting of either the leaders of organizations or the chairs of the interest-group committees or task forces dealing with the planning review. This group had representatives from 21 different organizations, including provincial agencies. It remained in place for the remainder of CPDR's work. The Leaders Group provided a forum for CPDR to announce details of schedules, bring issues of common concern to the table, and test options. It gave organization leaders an opportunity to hear each other's positions and to discuss them in a nonpublic forum.

Most organizations invited CPDR to take part in their annual conferences or workshops. Some, such as the Ontario Professional Planners Institute, organized a series of regional meetings so that members could talk directly with CPDR. Several organizations, such as the Urban Development Institute, Canadian Environmental Law Association, and Ontario Professional Planners Institute, attended CPDR press conferences and issued their own press releases on CPDR's work.

In addition to organized interest groups, the general public wrote submissions and participated in public forums and meetings. Communication links were provided through radio broadcasts, phone-in programs, other media — such as television and newspapers — and a 1-800 telephone number.

Involvement of decision-makers

In this planning-reform process, three groups of formal decision-makers were important: provincial politicians, municipal politicians, and provincial bureaucrats. Provincial politicians were important because final approval of policies was in their hands. Municipal politicians were important because they ultimately carried significant responsibility in implementing policy and also because they constituted an important lobby group — strong objection to policy by this group could mean new policies would not be adopted. The provincial bureaucrats were important because they made recommendations to their respective ministers on the content of recommended policies and were responsible for administration of policy implementation.

Provincial politicians were difficult to involve because of the limited time they were available. Information distributed by CPDR was sent to all sitting Members of Provincial Parliament (MPPs) and to their constituency offices. Commissioners or representatives of CPDR met twice with the caucuses of the opposition parties and three times with the government caucus before submitting its final report.

CPDR met regularly with the Minister of Municipal Affairs and with other ministers at their request. Regular meetings were held with representatives of Cabinet Office, and two meetings were held with a subgroup of Cabinet. Some sitting MPPs attended the public forums in their constituencies. After submitting its final report, CPDR met with individual ministers and the Premier between June and August 1993 to review proposed changes.

In the June 1992 interim report to the Minister of Municipal Affairs, CPDR asked the government to regard the review of the proposed policies in the draft report as fulfilment of the requirement for review of policies under the *Planning Act*. This would allow the government to approve some or all of the policies in the final report without further consultation. The request had the effect of asking the Minister of Municipal Affairs to show support for the idea of a comprehensive set of policies while CPDR was still refining policy proposals. The affirmative response by government led to significantly increased interest in CPDR's work. CPDR received more than 1 200 submissions on the draft report. This step also established an implicit agreement from provincial politicians that comprehensive policies were a useful approach to resolving some concerns about the planning system.

Municipal politicians were easier to involve in the process. Some participated in working groups. Through the Association of Municipalities of Ontario (AMO), the municipalities had a task force to deal directly with CPDR. The chair of this task force also sat on the Leaders Group. AMO invited the commissioners to several local and provincial conference sessions on planning reform.

CPDR arranged meetings with local and regional or county politicians in public-forum venues before the public forums. Some politicians also participated in the forums by making presentations or by observing the proceedings. Local politicians sent individual submissions; municipalities sent submissions; and AMO submitted briefs through its task force. AMO representatives also attended all of CPDR's press conferences.

Insofar as provincial bureaucrats were involved, the staff of relevant provincial ministries and agencies provided support through initial briefings and information and by inviting CPDR to internal committee meetings. For example, CPDR met with a committee of deputy ministers from seven key ministries and with an interministry land- and water-use committee at each stage of the work. These were briefings and provided opportunities to discuss concerns and answer questions.

CPDR also established its own interministry group of provincial officials, with representatives from 13 ministries. As issues emerged individual ministries arranged working sessions that included both central office and field staff. This happened most often with ministries of Municipal Affairs, Environment, and

Energy and Natural Resources. While conducting public forums CPDR visited several regional offices of provincial ministries.

The province's decision to consider the consultation on the draft-report policies as fulfilment of the requirement for consultation under the *Planning Act* meant that consultation on the draft report became a joint CPDR and provincial consultation. At the bureaucratic level, the province formed an interministry policy committee to consider provincial concerns and to provide ideas on the policies that were going into the draft report. This group also remained in place to address CPDR's final recommendations on behalf of the government, thus reducing the time required to review and implement policy recommendations.

Resolution of conflicts

A key to reaching common ground on CPDR's recommendations for new policies and implementation procedures was the general dissatisfaction with the existing system. Concerns varied. For example, the development industry was interested in a more timely process, whereas environmental interest groups wanted to improve environmental protection.

Several aspects of the process helped the participants reach agreements. First, CPDR did not set out to reach a consensus. The terminology used was *common ground*, in recognition that total agreement might be impossible but that even with some level of disagreement, a political decision could be made. This approach meant that one interest group could not stop the process by saying their concerns were inadequately addressed.

Second, the process was very open. The presentations from different interest groups were available through the public forums and through public access to submissions. Open press conferences allowed the media direct access to the interest groups and their positions, which could then be challenged by other interest groups and the public. This openness allowed interest groups to test the public acceptability of their viewpoints. Positions that were unfavourably received were generally amended.

Although the process was open, the Leaders Group and working groups also allowed various interest groups to explore their differences and areas of agreement in a private forum. Members commented that otherwise they rarely had the opportunity to discuss their concerns with other stakeholders in a non-threatening way. This approach helped to build understanding and trust among the participants, who, as influential members of their interest groups, exported this trust and relationship-building to their organizations.

The process was highly iterative, and this, too, was important to resolving conflicts. Ideas could be proposed and tested without their proponents' asking for a firm commitment, and agreement was built slowly and incrementally. Relevant information could be assembled to inform participants about issues and options. This process allowed participants to learn and to make the internal adjustments needed to convey their understanding and support to their various organizations.

Separate, closed meetings with interest-group task forces and committees allowed for frank discussions with CPDR. Concerns and frustrations could be expressed without the scrutiny of the public and the media. These sessions also allowed CPDR to challenge interest groups and to test their positions, new proposals, and options. These meetings helped build a level of trust and a relationship between CPDR and interest groups. Stakeholders began to understand that their interests were not being ignored and that innovative solutions were needed to reach a common ground.

Finally, CPDR continually searched for common interests. For example, in the working groups it became apparent that the development industry did not oppose environmental protection but did wish to have clear rules about what was to be protected and fair treatment during the transition to a new system. Similarly, environmental interests were not opposed to a more timely process. If policy could address their concerns, they could be relieved of some of their watchdog responsibilities and the significant personal costs involved. Both groups had a common interest in good policies.

Outcomes and impacts on policy design and implementation

The key outcome of this process was approval of a set of comprehensive policies. Most stakeholders saw these policies as beneficial. The provincial government and bureaucrats saw policy as essential to protecting public interest. The development industry, although not unanimously, perceived policy as a way of clarifying the rules they would have to recognize. The industry anticipated reduced delays in approval processes if decision-makers respected the policies. Municipal politicians, again not unanimously, saw increased emphasis on provincial policies as a trade-off for gaining more local control over specific development decisions, although some doubted that delegation of approval powers would follow. This scepticism was reinforced by the release of the extensive implementation guidelines, which seemed to represent further government red tape and an infringement on municipal decision-making powers. Some municipalities saw provincial policies as giving support to what they were already attempting to achieve and welcomed the new policies. Citizen and environmental-interest groups saw policy as a safeguard against municipal governments' making short-term, politically expedient decisions.

A second outcome was to demonstrate the use of a participatory decision-making model by a provincial commission. Use of a commission — an arm's-length agency — allowed a focused debate to take place on public interest in issues in a way that avoided parochial ministry or interest-group positions. It allowed all sides to have a voice in the process. As a result of the working-group structures, various interests could learn of about each other's views first hand and discuss their perspectives with each other in a private forum. This had the effect of building relationships that in the future might help in implementing the policies and in resolving new issues of common interest.

The government extended the working-group model to the implementation process, establishing stakeholder working groups to help develop the implementation guidelines. The direct use of multistakeholder groups was an innovation at this stage of the process, which had previously been left in the hands of the bureaucrats.

The framework for policy development was intended to be strategic and flexible. Even in this context, however, two vital aspects of the process evolved that were different from or more important than anticipated. First, the provincial government decided to consider the consultation on the policies in the draft report as fulfilment of the requirement for consultation on provincial policy, under Section 3 of the *Planning Act.* This idea emerged after it became clear that people were in general agreement about the need for policy and gave considerable support to the draft polices circulated in the April 1992 newsletter. CPDR's rationale was that endless consultation would not be constructive. Government support for CPDR's request brought attention to, and an increased engagement in, the CPDR process of policy development. It also forced the ministries to clearly define their policy interests.

Second, the media coverage and interest-group newsletters and publications became an asset to the process. CPDR had a communication plan, developed by a communication consultant, that involved the media in a substantial way. Nevertheless, the extent of positive response from the media was unanticipated. They actively sought out CPDR for interviews and participation in media programs. This allowed CPDR to have greater access to the public than would have otherwise been possible.

In addition, for many articles, the media selected general issues identified in the policy process and investigated these in the local context. This provided the public with information that CPDR had neither the time nor the resources to develop.

Research

CPDR gathered research with the help of consultants, a full-time researcher, a full-time librarian, staff of provincial ministries, and participants. Written submissions were analyzed by CPDR staff. The research was specifically aimed at supporting the process of policy development. Basic research on fundamental issues of policy (for example, water quality) was not part of this process. Policies were developed in the context of existing knowledge and professional experience. Although this might have seemed to be a limitation, it was apparent that there was a considerable gap between existing knowledge and its application.

A second aspect of research was the issue of whether full knowledge of how to address problems was needed before policy approval. CPDR's position was that establishing planning goals or policies without absolute clarity on how those ends could be achieved was not only feasible but desirable. Resources, including research, could then be organized to respond to that goal. However, given the political nature of policy development and the voice and power of stakeholders, CPDR recognized that its ability to set goals well beyond the confines of current knowledge was limited. The objective of research was to ensure that the commissioners and staff understood the current information on a topic and that the information was incorporated into the process.

For example, one set of issues — specifically, time delays, conflicts leading to OMB hearings, and related matters — needed better documentation. CPDR hired a consultant to study the actual operational realities of the planning system. This took 6 months, cost about $80 000, and involved a series of 26 municipal case studies in various planning contexts throughout the province. The studies collected data on planning activities; as well, planning officials were asked to give their opinion about problems in their jurisdictions and to suggest solutions.

A legal consultant was hired in the fall of 1992 to review the legal implications of the terminology used for requirements in legislation to implement policy. One issue was whether *have regard to, conform to*, or *be consistent with* should be used in policy statements. This was a short-term contract, costing about $5 000.

The full-time staff researcher investigated issues and concerns that arose during the policy-development process. Usually, the researcher consulted secondary data, professional reports, journals, government publications, or other sources of relevant information; key-informant surveys were also used to inventory both opinions and experience, as well as to suggest sources of information. In total, the researcher prepared 114 written research briefs for the commissioners and staff. In at least as many cases, the researcher presented verbal briefs on smaller research items. The researcher also built the glossary of terms used for policies and for the final report (Moull 1993). The cost for the researcher's services, including salary and benefits, was about $160 000.

The full-time librarian was engaged to collect and organize relevant information and to find the documents requested by the commissioners, the researcher, and staff. The librarian recorded the title and author and prepared a brief description for each document as it was received. By the end of the commission's work, the library contained more than 6 000 items.

At the start, the commissioners were briefed by officials from relevant provincial ministries and agencies on information related to policy and planning reform, as well as on their mandate, roles, and activities. Typically, staff of key ministries met several times with CPDR over the course of the work to provide updates and to respond to CPDR's initiatives. CPDR received both opinion and information as part of the consultation process. Municipal studies, reports, individual research reports, and papers were also presented at public forums or sent to CPDR.

The internal approach to research was one of collaboration and mutual assistance. CPDR drew on staff, as necessary, to provide support for the process, and the staff occasionally asked the commissioners to help in operational matters, such as packaging reports and newsletters. Mutual respect and a team attitude were highly valued. When time permitted, the commissioners would take on research tasks directly, usually on topics related to policy and institutional change. For example, a commissioner prepared the background research on roles of and relationships between First Nations and municipalities. The staff researcher helped by collecting data on the number and location of reserves, land claims, and similar information.

CPDR undertook little direct training other than development of skills in the use of computer software. The researcher and staff went to a limited number of conferences and seminars on planning and policy issues and research methods used for the work; for instance, they attended a meeting of the Urban and Regional Information Systems Association. All staff attended some of CPDR's scheduled events, such as working-group sessions, public forums, and meetings with committees, to gain insight into the process and have first-hand experience of how the information was used. The researcher also attended most briefing sessions with ministry staff.

Research methods

The commissioners, along with the researcher, executive director, and communication consultant, established the research agenda. Agreements on research priorities, schedules, and implementation were reached by consensus at weekly staff meetings. Occasionally, initial research uncovered either substantive or methodological issues that resulted in further discussion by the group.

Information and data were usually collected from secondary sources, although some primary research was conducted that involved data analysis (for example, defining municipal planning costs as a component of municipal expenditures). Typically, the research was oriented to gaining an understanding of practices, standards, and procedures used in the planning system.

Although much of the research focused on procedural concerns, work was also done on some substantive issues related to the environment, land stewardship, and resources. Administrative matters were the focus of the economic research. Social concerns were addressed through specific issues, such as housing, public transit, and servicing. This research was straightforward — no innovative or complex methodologies were used.

Although procedural and environmental issues still dominated the agenda, the consultation process created an important balance of opinion and information. Personal experiences, concerning such matters as the costs of development, financial issues in farming, illegal rental units, and social housing, were presented in this context. This information was retained in the process through the commissioners' reading of all submissions. As well, summaries of comments and suggestions about policy or other recommendations were prepared and made available to the commissioners as they revised their recommendations.

CPDR's orientation to public involvement helped to ensure the rigour of the research. Through the newsletters and public forums and discussions, most research results became part of the public debate, and the participants effectively provided a check on accuracy and completeness. Any error of omission or content became a point of criticism of CPDR, making it in the commission's interest to avoid these errors.

Roles of disciplines, institutions, and organizations

The commissioners and staff had qualifications as lawyers, engineer, geographer, planners, and agronomist. The commissioners and most staff also had experience in applied research.

CPDR had no direct link with other organizations. A number of organizations were consulted for information and comments on proposals but played no direct role in setting the research agenda, carrying out the research, or developing recommendations. Several individual researchers approached CPDR with proposals of their own. These research proposals tended to focus on further details about the current situation or innovation in the system. CPDR funded none of these proposals, because it felt that problems in the system were adequately documented for the commission's purpose.

Although several individual academics from professional and related programs followed CPDR's work, sending submissions and presenting these at public forums, academic and research organizations did not engage in this process. CPDR reached out to this sector by visiting several research institutions and schools and providing information through seminars and classes. As well, CPDR had the relevant academic programs placed on its mailing list. However, these mailings generated little interest or response, and it is unclear why academic and research organizations oriented to public policy and planning took little interest in becoming involved in CPDR's work.

Research and policy links

In terms of policy development, CPDR's research made three main contributions:

+ It helped to set the context. Before and during CPDR's work, considerable research was done that focused on various planning concerns, such as watershed planning studies, urban-transportation studies, comparative studies of planning systems and policy approaches, and the work of the Royal Commission on the Future of the Toronto Waterfront. This information raised awareness about issues and provided priorities for policy development. The importance of this body of research to the decision to appoint a commission and to the subsequent success of the policy development is impossible to precisely define, but clearly it helped to politicize planning concerns.

+ It played an obvious role in the development and approval of recommendations and policies. Policies had to be designed primarily on the basis of the large body of information that was then available.

+ It informed the policy implementation. However, although some information and methods are available to address emerging issues — such as cumulative impacts and the ecological significance of specific natural features — both information and techniques must be improved if high expectations about the usefulness of the policies are to be met. Ongoing research efforts will be needed to address both the anticipated and the unforeseen issues that emerge as attempts are made to implement and monitor the new policies.

Background studies involving more than one stakeholder (such as joint municipal–provincial or private-sector–public-sector studies) seemed to carry the most credibility with interest groups. Such research had the effect of bridging institutional and organizational differences.

Concluding observations and links with theory

Characteristics of the process

Several key characteristics of the process of developing integrated policy emerged from the case study:

+ The political, social, economic, and information contexts for policy development were opportunistic. A need for policy had been recognized for more than 20 years. The province had initiated planning initiatives (such as that for the Oak Ridges Moraine), in part because of a lack of policy, and had then encountered considerable difficulty because of the lack of a policy framework. Finally, a new government was in place and was interested in making changes to the planning system.

+ The use of an independent Commission of Enquiry allowed debate on policy proposals and related amendments to occur outside the context of partisan politics. This approach meant that the focus could be kept on the substance of policies.

+ The selection of the CPDR chair established the expectation of change. The chair was an individual well known to the public and to politicians and had acted as a champion of change and provincial policy.

+ CPDR established a public agenda for the review and placed policy development on that agenda. A clear schedule helped CPDR and the interest groups anticipate developments in the process and organize their resources accordingly.

+ CPDR set out to recommend policies that would address concerns and had enough support that the government would adopt them. CPDR did not set out to recommend ideal policies based on normative research. Because of this, individual stakeholders recognized that specific details or concerns had to be traded off to accommodate other interests.

+ CPDR aimed to emphasize general principles and values in policy content. In some cases, these values precluded any change (for example, no development on significant wetlands), but often they encouraged change (for example, affordable housing). It had to be recognized that in specific contexts, these values might conflict.

+ Integration of policies and values is not possible at the provincial level. Policy details and conflicts were seen as being best resolved at local and regional levels and by OMB. Integration would occur through actions, rather than through the specific content of policies.

+ The process used by CPDR was inclusive and iterative. An iterative learning process takes considerable time and energy from all the participants. Those who wished to participate fully had the opportunity to do so. Several types of forums and other mechanisms were available to allow stakeholders to participate. The process provided the opportunity for individuals and organizations to explore options and learn along with CPDR.

+ Stakeholders were active participants in the process. Ideas for policy and other proposed changes were generated by many people involved in the planning system.

+ Research generated information needed to resolve concerns as they emerged. This approach assumed that sufficient information and experience to develop and implement recommendations and appropriate policy was either in the system or could be developed with some effort. Research was integrated into the process of formulating policy, finding specific details on issues, developing options for solutions, and approving and implementing policy.

+ Implementation was part of the policy-development process. A major component of CPDR's work and recommendations dealt with changes to the system to make it more functional and to ensure that policy would in fact be implemented; any proposed solution had to be seen as being feasible to implement. Shifting authority to local levels was a significant related thrust in the recommendations.

+ The public process was open, timely, and efficient. CPDR's review of the planning system, including development of policies, took less than 2 years. Government review of the recommendations and approval of new legislation and policy took an additional year.

Additional observations

To validate these observations, Marshall et al. (1995) undertook a survey of 13 key informants. These informants were directly involved in CPDR's review process; the process used by the government to review, amend, and approve the recommendations; or both. Respondents represented interest groups, government, and CPDR. They generally agreed the process had the characteristics as described above. However, two areas of difference or concern emerged from the survey.

The first of these was the extent to which the comprehensive set of policies was integrated. Respondents generally agreed that in terms of scope and integration, this set of policies was less than ideal; in particular, the policies were weaker in social and economic terms than in environmental ones. One possible explanation for this is that we still lack a good understanding of genuinely well-integrated policies. Another is that the mandate of CPDR's work and of the *Planning Act* emphasized physical and environmental considerations, so the work inevitably focused more strongly on these matters.

The second of these concerns was about the linkage between CPDR's work and the government review and approval of its recommendations. The review process was transparent and accessible under CPDR's management. However, once the final report was submitted, the Ministry of Municipal Affairs managed the review and approval process and the development of implementation guidelines, and respondents felt that this stage of the process was less open and accessible. The result, particularly in the view of interest groups, was that policies and proposals for legislative change began to reflect the interests of the government and bureaucracy, rather than of the stakeholders.

The role that CPDR played in this final stage was that of independent critic of the government's proposals. The former commissioners made comments in the media, at conference presentations, and before the standing committee of the legislature that reviewed the draft legislation. These actions were alienating to the bureaucracy, but they failed to go far enough in the view of the interest groups. A closer relationship was obviously needed between the policy-building process and the formal policy-approval process.

Related theoretical models

CPDR's review process does not fit the traditional model of comprehensive planning and policy analysis, which is based on the assumptions that objective methods can and should be used to make policy decisions more rational; that rational decisions materially improve the problem-solving ability of organizations; and that management and decision-making systems are comprehensible in terms of inputs, outputs, their environment, and feedback loops (Friedman 1987). In this traditional model, used in a planning context, analysis and research directly inform policy; the state exercises considerable authority; and the outcomes are plans or policies that are effectively regulations in their degree of detail.

CPDR's process differed from this model for two reasons: one is philosophical; the other, practical. First, land-use planning and planning policy affect private as well as public lands. The state has control over public resources and public land, and state-derived policy tied to state management systems and based on a traditional model may indeed be feasible for the sustainable management of these resources. However, on private lands, change is initiated by private landowners. In a democratic system, imposed state control over some of these changes may not be legal, but even when it is legal, negative public reaction to such imposition can result in poor implementation, a change in government, or both. In these situations, "ownership" of a policy must be shared between the state and the private sector. The philosophy underlying effective planning policies, then, is to actively seek community engagement in the process, rather than merely relying on a good analytic rationale.

The second reason is practical. If the state developed and approved policies unilaterally, it would require a large bureaucratic infrastructure to enforce implementation. In Ontario, deficit budgets and high debt have forced substantial restructuring in the public service, which is reflected in the Conservative government's decision to further amend the legislation and remove provincial approval. The province simply doesn't have the resources to monitor and enforce implementation of planning policy. Furthermore, considerable authority in planning had already been delegated to regional municipal governments. Their cooperation would be needed for implementing any policy.

CPDR's model has its roots in the idea of strategic planning. In this model, policies constitute a framework of aims. A common philosophy becomes the basis for action. Organizational and interorganizational complexity is assumed. Investigating facts, clarifying values, and building working relationships are all considered components of decision-making. Uncertainty and a complex relationship between the technical and political aspects of the decision-making process prevail. The approach to making choices is synoptic, elaborative, interactive, accommodative, and decisive. The linearity, objectivity, certainty, and comprehensiveness of traditional approaches are replaced by cyclicity, subjectivity, uncertainty, and selectivity (Friend and Hickling 1987).

In their book, *Leadership for the Common Good*, Bryson and Crosbie (1992) emphasized the complexities of shared-power relationships in formulating and implementing public policy. They identified key tasks in the policy-change system:

+ Understanding the social, political, and economic givens;

+ Understanding the people involved, especially oneself;

+ Building teams;

+ Nurturing effective and humane organizations, interorganizational networks, and communities;

+ Creating and communicating meaning;

+ Making and implementing legislative, executive, and administrative policy decisions;

+ Sanctioning conduct — that is, enforcing constitutions, laws, and norms and resolving residual conflicts; and

+ Putting it all together (actions and outcomes that are noticeably better).

These concepts fit well with the consultative approach used by CPDR, but Bryson and Crosbie did not discuss the important role of research. Information is explicitly or implicitly assumed to be part of the process, but strategies for generating information are not clear.

Strategic approaches to policy development are most clearly reflected in ideas about information in the literature on soft-systems methods. This approach emphasizes the search for meaning, rather than analytically generating integrated solutions. Information is based on experience and action, and integration is part of a social-learning process (Checkland and Scholes 1990). This puts research into the policy-making process, as part of the search for meaning. Some theoretical approaches to policy formulation and research can therefore be said to generally fit the case study. Although CPDR did not explicitly draw on this literature in formulating its concepts, in practice it shared a philosophical common ground and similar methods of operation.

Implications

The CPDR process is a model that could be used for policy development in other jurisdictions. However, its general applicability has several limitations. First, the idea of integrated policy conflicts with the reality of political and administrative structures. Getting political agreement on a mandate to create integrated policies is difficult. Moreover, the scope of policies and the extent of integration are limited by the structures government uses to create and implement them.

Second, good process and information act only as partial buffers to political and administrative agendas and priorities. In this case, representatives of government agencies and elected officials were included in the process. However, once in charge of the formal approval process, the political administrative system made changes that conflicted with some agreements and understandings that had developed through CPDR's work. There was no obvious way to strengthen this linkage.

Third, this is a resource-intensive process. Although CPDR's budget was reasonable, the time spent by public and private interest groups would be unaccounted for in a costing based only on public expenditures. This is a real expense that increases the total cost several times over the actual dollars spent on CPDR.

Fourth, this is a professional process. In this case, most interest groups used planners or lawyers to prepare or present submissions and comments. These individuals were generally familiar with planning issues and related studies, and this level of expertise allowed the process to proceed relatively quickly, with limited effort in basic education and research.

Finally, this process came out of a tradition of planning and a history of review of the system going back at least 20 years. This context was important to CPDR's success but may not exist in other jurisdictions. Also, a significant body of relevant information and research was available; the chair was exceptionally well respected; and the sociopolitical environment was pluralistic, with an established participatory tradition. The media were generally supportive, and — perhaps most significant — the process took place in a context of general dissatisfaction: no change was not an option.

Although there is much that might be learned from this case study, every policy process has a unique history and context. This uniqueness must be understood and respected to ensure the success of such policy exercises.

PART III

SOUTHERN PERSPECTIVES

Chapter 6

Wetlands Management in Ghana

T. Anderson

❖ ✦ ❖

Introduction

Wetlands in Ghana are unique ecosystems that provide valuable products and services to satisfy social, economic, and ecological needs at the local, national, and international levels. Ghana's wetlands support fisheries, play an important role in flood assimilation, and provide a source of food, medicines, fuel, and building materials for local people. Nevertheless, past policies aimed at wetlands management were heavily tilted in favour of industrial use, ignoring the contributions of wetlands to local livelihoods and, in many instances, causing negative environmental impacts on the broader wetlands ecosystems.

This case study draws on research on wetlands management in Ghana — a project undertaken by Friends of the Earth, Ghana (FOE–Ghana), a nonprofit, nongovernmental organization (NGO) contributing to the ongoing process of integrated-policy development in Ghana. This chapter presents and explains the key social, economic, and ecological importance of Ghana's wetlands and provides an overview of the policy context in which the management of wetlands in Ghana is currently being addressed. The study focuses specifically on the following three areas of interest to the development of integrated policy:

+ The design of research aimed at integrated policies;

+ Factors that facilitate or constrain the use of integrated policy; and

+ The development of tools, methods, decision-making processes, and institutional arrangements to support the design of research and policies to integrate environmental, social, and economic concerns.

Drawing on FOE–Ghana's research, this chapter offers recommendations for the design of an effective integrated-policy mechanism and institutional framework for the sustainable management of wetlands in Ghana. Finally, this chapter offers some recommendations for developing broad policy integration for planning in Ghana as a whole.

Ghana's experience in integrated-policy development

Ghana's experience in planning for development dates back to the colonial era. Since independence, successive governments in Ghana have attempted to promote development by designing national development plans. A review by Trevallin (1994), commissioned by the Government of Ghana, examined how integrated policy had been designed and implemented within the national development plans. Trevallion identified the need to integrate economic, social, environmental, and political concerns into development planning and policy. He came to this conclusion based on his assessment that the spatial structure of the economies of most developing countries (including Ghana) was characterized by the following interconnected problems:

+ *Urban primacy* — Too much concentration of resources in urban development and capital cities, resulting in a lack of resources for the rural areas;

+ *Rural–urban migration* — A large influx of rural population to urban centres, stressing urban infrastructure beyond its carrying capacity; and

+ *Environmental degradation* — Deterioration of the natural environment and destruction of natural resources, thereby reducing the stable functioning of ecosystems.

In effect, the structure of developing-country economies has in the past had a negative effect on rural livelihoods, which depend heavily on agriculture and natural resources.

Similarly, Mamphey and Agyei (1985), in their study of planning and management of human settlements, suggested that to enhance the productive role of rural communities at all levels, the following problems needed to be addressed:

+ Monolithic and overcentralized planning for development, without feedback from the local level;

+ The disparate and nonintegrated nature of national and local development proposals, both thematically and geographically;

+ Failure to detect and harness available human resources and abilities in rural areas;

+ Failure to understand the social structure of rural people and its impact on macropolicy implementation;

+ Overemphasis on development that directly or indirectly stimulates urban growth (that is, of primary cities), thus diverting human and financial resources from the transformation of natural resources that promotes the growth of local economies;

+ Overemphasis on the service sector, to the disadvantage of investment in direct-production activities; and

+ Inattention to local-level development and the aspirations of local people in the formulation of broad policies, resulting in development priorities that alienate local communities.

These studies presented broad criticisms of the failure of Ghanaian political and economic structures to address the immediate needs of rural people and their reliance on a sustainable natural-resource base. Within this context, planners and policymakers are seeking a new approach for sustainable development in Ghana.

New development perspectives

FOE–Ghana has been involved in the development process in Ghana in recent years by studying and contributing to the discussions of the 25-year development policy framework known as the Vision 20/20 project (see NDPC 1995). This new policy guideline recognizes fundamental relationships among environmental degradation, the distribution of human population, economic activity, and the pattern and scale of human development. The guideline also recognizes that policy failure at the local and regional levels inevitably means policy failure at the national level.

In this regard, development-oriented action at the local level must meet the socioeconomic aspirations of the community within the regional and national organizational structures. This is important because the success of national socioeconomic and spatial policies depends on complex interrelationships and activities at the local level carried out voluntarily by local communities and the general population as a whole.

To promote change that addresses these issues, the Government of Ghana put into place a strong local government law to enforce integrated development at the grass-roots level. This is popularly known as the "local government law on decentralization."

Decentralization is being implemented to increase local participation in development and management. Some political and financial power has therefore been devolved to the district councils to enable them to contribute both to top-down and to bottom-up policies and programs.

The experience so far indicates that

+ National guidelines based on indigenous political philosophy are indispensable to the development of rural economies; and

+ Rural resources should be considered in the context of ecological opportunities and constraints.

Review of case studies of integrated policies

To learn about the African perspective and contribution to integrated research and policies, the International Development Research Centre (IDRC) held a workshop in Abidjan, Côte d'Ivoire, in 1994, at which papers on the topic were presented. African research institutions, including the Nigerian Institute of Social and Economic Research, Kenya's Department of Land Development, and the Department of Sociology at the University of Cape Coast in Ghana, participated, along with the Policy and Planning Group of IDRC.

At the workshop, an important distinction was made between *interdisciplinary* and *multidisciplinary* research. Interdisciplinary research seeks to establish a common problem and then common knowledge that leads to the development of a common theoretical link among disciplines. Multidisciplinary research promotes the use of a variety of disciplinary perspectives to tackle a common problem. Whereas interdisciplinary research can easily lead to disciplines interlinked in a project, this is not true of multidisciplinary research. Integrated research is therefore more than the sum total of research carried out by a variety of distinct disciplines; synthesized knowledge and a common frame of reference derived from various disciplines are required for pursuing a given problem and developing a unitary solution.

A number of key observations and recommendations regarding the practice of integrated research emerged from the workshop:

+ Essential ingredients for formulating integrated rural policy include ethnoscientific objectives, involving the development and use of indigenous systems of knowledge, as a basis for development planning.

+ Integrated-policy needs should be dealt with at all stages in research (that is, in problem identification, team selection, data collection and analysis, synthesis, and presentation of findings).

+ To encourage interdisciplinary research, research institutions must acknowledge that it needs to be supported through sustained capacity-building, training, and reward systems distinct from current systems that promote unidisciplinary excellence. (This also means sensitizing policymakers and resource allocators to the need for the interdisciplinary approach.)

- Integrated social-science research needs to adopt the principle of going out to the people, learning with them to identify problems and then finding solutions together, using a blend of qualitative and quantitative data and analysis to discover what the perceived problems, needs, and priorities are.

- Integrated methodologies and theoretical frameworks should concentrate on areas that enhance the integration of social-science research methodologies.

Wetlands and the National Environmental Action Plan

In the early 1990s, the National Environmental Action Plan of Ghana (NEAP) was launched under the auspices of the then Environmental Protection Council. The program was sponsored by the World Bank and the United Nations Environment Programme. NEAP's objective was to identify areas of environmental concern to the government and people of Ghana and to prioritize key areas for action (Laing 1994). NEAP defined the scope for environmental intervention in Ghana and set out the policy and strategy for managing its environmental resources. This sustainable-development approach provided a coherent technical, institutional, and legal framework for intervention. NEAP covered all aspects of natural-resource use in the country, including wetlands ecosystems. This plan was expected to form the basis for addressing priorities within the environmental sector. NEAP identified the following policy actions for marine and coastal ecosystems:

- Adoption of a fisheries-management policy;

- Adoption of proposed legislation and regulations on coastal-zone management; and

- Establishment of protected areas in coastal wetlands.

For implementation, NEAP suggested the development of appropriate mechanisms to

- Improve the scientific base of environmental policy, through research programs;

- Assess the potential impacts of certain public and private projects on the environment and integrate the environmental dimension in natural-resource policies;

- Establish and implement appropriate standards and guidelines to ensure an acceptable level of public health and environmental protection;

- Harmonize appropriate legal instruments; and

- Improve access to information on the environment.

NEAP identified the following research topics to fill gaps in the data on wetlands ecosystems (lagoons, marshes, mangroves, and estuaries):

+ Inventory of all wetlands;

+ Physicochemical and biological character of the wetlands;

+ Socioeconomic and cultural importance of the wetlands; and

+ Current wetlands-management practices.

The most important research objective, according to NEAP, was to identify and select wetlands for protection, based on the following criteria:

+ Fragile wetlands, that is, those vulnerable to irreversible change;

+ Wetlands with economic value, for example, for tourism, sediment traps, and fish-breeding grounds;

+ Wetlands with traditional, cultural, or social significance; and

+ Wetlands of international importance as habitats for migratory birds.

Ghana is a signatory to eight international conventions relating to coastal protection. Ghana's legal regime for coastal-zone protection is adequate for environmental needs. NEAP did not suggest any additional studies but implied that coastal-protection works would be environmentally beneficial to coastal wetlands and that there was a need to identify and establish areas for protection and for sustained resource management.

Allocation of responsibility for wetlands management

NEAP's allocation of responsibilities for protection of wetlands ecosystems was as follows:

Responsibility	Institution(s)
Data gathering and monitoring	Wildlife Department, Forestry Department, Fisheries Research, Institute of Aquatic Biology, The universities, Architectural and Engineering Services Corporation, Water Resources Research Institute
Standard setting	Wildlife Department, in consultation with Forestry Commission, Mineral Commission, Lands Commission, Fisheries Department, EPC
Legislative enactment	Ministry of Lands and Natural Resources, with advice from EPC
Legislative enforcement	Forestry Department, Wildlife Department, Fisheries Department, district and metropolitan assemblies, civil-defence organizations, Ghana's police service, Ghana's navy
Execution of environmental projects and programs	Supervised by EPC, in consultation with Forestry Department, Fisheries Department, Mineral Commission, Lands Commission, Wildlife Department

Note: EPC, Environmental Protection Council.

Lead agencies for implementing NEAP

According to the institutional framework outlined in NEAP, the following institutions act as lead agencies:

+ Environmental Protection Council (EPC);

+ University of Ghana (Geography and Zoology);

+ Department of Wildlife (Ministry of Lands and Forestry);

+ Forestry Department (Ministry of Lands and Forestry);

+ Institute of Aquatic Biology of the Council for Scientific and Industrial Research; and

+ District and metropolitan assemblies.

However, an assessment of the key institutional framework revealed the absence of three important stakeholders: the private sector, NGOs, and local communities.

Case study of wetlands management in Ghana

Before this study, little documented information was available on wetlands management in Ghana. A survey of the water surface in the country in 1986, not including Lake Volta, indicated that Ghana had about 73 000 ha of water surface after the rainy season. Swamps occupied 34 000 ha, mostly in the Northern Region (Kapetsky 1991). Most of these may be described as wetlands, although a great many of them likely dry out during the dry season. Local perceptions of the wetlands varied considerably, depending on the use and benefits enjoyed by the communities that depended on the resources. However, many local people may regard the wetlands as wastelands. People usually fish in wetlands with permanent water bodies. In coastal lagoons, salt-winning may be a major activity in the dry season. Traditional uses of the wetlands often tend to be exploitative, controlled only by taboo systems of management. In taboo systems, the people adopt totems, declare closed and open seasons, and may dedicate lagoons and lands to gods, making these areas sacred. Ownership of these resources may be based on tribal, clan, or family affiliations, depending on specific cultures and societies (Agyepong 1993).

The wetlands of Ghana are of two types — the coastal-zone wetlands and the noncoastal, inland wetlands — and can be further classified by water regimes, topography, soils, vegetation, animal life, and resource potential and use.

In 1992, with financial support from IDRC, FOE–Ghana undertook a study of wetlands management to satisfy NEAP requirements. The study sought to identify policy-related issues and provide management plans for the rehabilitation and conservation of wetlands in Ghana. The idea was to examine, from the viewpoints of diverse disciplines, the current status of wetlands management in Ghana, drawing from the experiences of the various stakeholders.

Research objectives

The research had seven objectives:

1. To prepare a National Wetlands Inventory (including an assessment and classification of all the wetlands according to their potential for enhancing socioeconomic development);

2. To identify and assess the causes and extent of damage or degradation to the wetlands, identify endangered species, and prepare a wetlands-degradation report;

3. To assess the socioeconomic conditions of the settlements around the wetlands (including demographic characteristics, sources and distribution of income, housing, education, and health services), examine the communities' perceptions of wetlands resources, and assess current institutional arrangements for managing the wetlands;

4. To provide a list of the communities successfully managing the wetlands, assess the core elements of their successes, and apply the experiences for replication and formulation of management plans for other wetlands;

5. To determine community responses to resource depletion (including socioeconomic shifts and lifestyle changes);

6. To design an action program for wetlands rehabilitation, with definitions of roles for each stakeholder; and

7. To develop policy guidelines for the management of wetlands resources (including specific guidelines for the protection of endangered species and wetlands) and prepare a wetlands adoption plan for NGOs (including guidelines for community involvement in sustainable wetlands management).

The research process

The research process and methods played an important role in bringing about integrated approaches to problem-solving for wetlands management. The research methodology was largely participatory, cost effective, and rapid. The data collection was designed not only to gather objective facts but also to facilitate an understanding of the processes going on behind the observed facts. This facilitated interaction with traditional and other leaders and the citizens within the immediate vicinity of the wetlands.

The process was designed to ensure the maximum participation of all relevant organizations and involved eight main steps:

1. *Establishing the research team* — A project coordinator and assistant formed the core of the research team. These were the director and the campaigns coordinator of FOE–Ghana. Other members of the research team were experts in socioeconomics, water-resources management, human ecography, and aquatic science. Other local experts and assistants were brought in when needed to assist in data collection. These included traditional leaders, who formed the local support team; extension officers; and various government departments, including Agriculture, Forestry, Fisheries, and EPC. Ten other field assistants were hired to assist in data collection, analysis, and discussions of wetlands policy in Ghana.

2. *Explaining the study to people in the relevant villages or communities around the wetlands* — A village-wide *durbar* was organized in the villages around the wetlands to brief the village leaders, residents, and local organizations on the proposed study and its objectives. (A community or village *durbar* is a traditional forum convened by the village chief to give leaders an opportunity to solicit a free and frank exchange on specific topics.) Questions from participants regarding the contributions they could make were answered, and the potential benefits of the research to the community were explained.

3. *Establishing local research support committees* — At the close of the village *durbar*, a local research support committee was established to assist the research team in carrying out the field study. This committee included local leaders, who could help arrange necessary appointments, lead discussions, identify contacts, and verify the data gathered from the village communities. The committee members were also interviewed regarding their roles as village leaders and how they could ensure that management plans emanating from the studies would be implemented.

4. *Collecting data and conducting field studies* — A search to identify available data and assess its usefulness to the study preceded the field work. Published and unpublished reports, local newspapers and newsletters, and topographical maps of all wetlands locations were reviewed. These sources allowed the research team to develop profiles — physical geographic characteristics, socioeconomic status, demographics, institutions, leadership structures, land-use patterns, and traditional resource-management systems — for each study site. Field data were collected during several visits by the research team, which used a combination of the following techniques:

   ✦ Direct observation and site visits;

✦ Formal (*durbars*) and informal discussions with individuals and groups, including special-interest groups (for example, women's groups and ecology clubs) and their members; and

✦ Household-survey questionnaires.

To collect basic demographic and socioeconomic data, the team developed a few short, structured questionnaires, which were administered to all heads of households. Special attention was paid to interviewing the formal and informal leaders of all the communities living around the wetlands, including village chiefs, queen mothers, councils of elders, and leaders of local institutions, such as churches, schools, and government committees.

5. *Analyzing, synthesizing, and verifying data* — The team reviewed and analyzed the data and information at each step of the survey, using the important findings to develop further questions to take back to the communities concerned.

6. *Verifying findings locally* — During the last few days of the survey, several *durbars* were held to share the data and preliminary research findings with the communities studied. These meetings also enabled the research team to verify the data collected and to strengthen the study with additional information to support its conclusions.

7. *Preparing and submitting the report* — The major elements accounting for the destruction of the wetlands and also elements that would facilitate the rehabilitation and conservation of the wetlands were discussed in the report.

8. *Preparing plans for wetlands resource management or rehabilitation* — A synthesis of all the lessons formed the basis of recommendations to government for designing policies and programs to enhance and promote self-help initiatives for grass-roots groups to manage wetlands resources and for preparing plans to enable the communities to manage wetlands resources, with the help of relevant organizations.

The research sites

The following categories of wetlands were chosen for the research:

✦ Wetlands located in areas strongly influenced by anthropogenic activities or in heavily urbanized and industrial zones;

✦ Wetlands located in lightly populated or rural areas;

✦ Wetlands earmarked for conservation as Ramsar sites under the Convention on Wetlands of International Importance especially as Waterfowl Habitat.

The team also made efforts to choose some sites with communities still engaged in some form of traditional wetlands management and some without. Based on these criteria, 14 wetlands sites were selected.

Modifications to the study

During the pilot study, the team realized that for development to be meaningful and sustainable, communities must develop a critical awareness of the causes of resource degradation and become involved in developing systematic ways of overcoming the constraints to sustainable development. There is a village or community belief that "no one knows everything and no one is totally ignorant." People have different perceptions, based on their own experiences. Development must, therefore, be pursued by enhancing real dialogue among the people and also between communities and the government, NGOs, and other stakeholders. The research team therefore used a listening-survey approach. This method involves observing and listening to people and discovering their concerns. Through this animation, communities are assisted in developing a critical awareness of their conditions of life and therefore their roles and responsibilities in their own development. Because effective and active participation of the communities is needed to achieve the objectives and overall goal of development, participatory dialogue and focus-group discussions were also used. The advantage of these approaches is that they prepare people to be involved in all phases of the planning process: problem identification, potential analysis, project selection, implementation, monitoring, and evaluation. Further information was obtained through field observation and physical inspection.

Integrating the research

FOE–Ghana consulted and involved all the important stakeholders and actors to ensure that the study took into account the full range of economic, social, political, and environmental perspectives, including those of EPC, the Forestry Department, the Game and Wildlife Department, academics, the Council for Scientific and Industrial Research, the district assemblies, local communities, industrial enterprises, government policymakers, and NGOs working in the sector.

What follows is a brief description of the institutions and personnel the research team approached for involvement under each objective of the research:

+ *Preliminary goal* — To reformulate the proposal design

    Issue — The redesign of project proposals to include a strong socio-economic emphasis and to reflect the framework of national priorities for action.

    Target institution: EPC — EPC had been the custodian of NEAP and was knowledgeable. FOE–Ghana felt that the involvement of this agency would ensure that the government took the recommendations seriously.

Personnel — The team discussed the research objectives with EPC. The team identified and approached senior officials and researchers and invited them to share in designing the strategy for the whole exercise.

✦ *Objective 1* — To prepare the National Wetlands Inventory

Issue — Identification, mapping, and documentation of all wetlands in Ghana (both coastal and noncoastal).

Target institutions: universities — The team chose the Department of Geography University of Ghana, as it was the seat of Ghana's first project on geographic information systems and had been involved in the coastal-management plans at the national level. The team also chose the Department of Zoology, University of Ghana, which had done some work on the scientific aspect of the Volta Basin project and its impacts.

Personnel — In an initial consultation, the team outlined the expected output and objectives of the research. The researchers targeted the personnel of the Geographic Information Systems unit and researchers from the departments of Geography and Resources; they had also been involved in the preparation of the NEAP on coastal-zone-management projects with the departments of Zoology and Oceanography. A process of dialogue was opened up with all those who had in one way or other been involved in wetlands studies, and these people agreed to assist in the coastal-wetlands inventory exercise.

✦ *Objective 2* — To prepare the wetlands-degradation report

Issue — Assessment of the state of wetlands in Ghana, using physico-chemical indicators.

Target institution: Institute of Aquatic Biology — The institute has been recognized as doing substantial work on the chemical analysis of the water bodies in Ghana, including work on pollution levels. NEAP identified the institute as the key institution for data gathering and monitoring.

Personnel — The team made contact with senior researchers at the institute who had done substantial work on the chemical state of a severely degraded urban lagoon. A series of discussions were held on the objectives and expected impacts of the study, as well as on future roles and collaboration between FOE–Ghana and the institute. This encouraged the institute's personnel to take part in the research.

✦ *Objectives 3–5* — To access socioeconomic factors affecting wetlands degradation, the ways the communities were managing the wetlands, and community responses to resource depletion

Issues — Assessment of the profiles of wetlands, identification of community economic status vis-à-vis wetlands degradation or conservation, and a study of how communities were adjusting to the depletion of resources.

Target institution: Ministry of Lands and Forestry — The ministry was responsible for major policy work on natural-resource and land use and was in a fairly good position to draw on various documents, both confidential and open, to enhance the output of the work. The implementation of the study's recommendations would also need the support of this ministry's Policy Planning Department. Another reason for involving the ministry was to avoid apportioning blame to the local communities. The team had already noted that the private sector was not involved in NEAP's roles and responsibility framework.

Personnel — The team invited the officer responsible for policy planning at the ministry, a socioeconomist, and a development planner to take part. The team selected a private consultant to support the socioeconomic profiles because the lead researcher was a development planner with a specialty in rural-energy policy, and an objective of the research was to examine energy exchange and the communities' responses to the degradation of the wetlands environment and its impacts on fuelwood supplies. The consultant's independence would also allow for constructive criticism of government or NGO positions that were not in line with the sustainable management of wetlands in Ghana.

◆ *Objectives 6 and 7* — To develop an action program and policy guidelines

Issue — Policy guidelines and strategies for rehabilitation

Target institutions: district assemblies — The team identified the district assemblies as the institutions responsible for prioritizing grass-roots development and ensuring effective and sustainable forms of community participation. For a policy to be well received by government, an insider at policy level in the Ministry of Lands and Natural Resources had to be brought into the picture.

Personnel — The team invited a member of the district assembly in Esuekyir, a dynamic teacher, to lead the community interviews and data collection. A socioeconomist and policy adviser at the ministry level was invited to help design the action program. A brief community workshop was organized to solicit views and support for the study. This exercise helped ensure that various groups felt comfortable expressing their concerns about wetlands management. It also engaged policymakers in constructive dialogue concerning the socioeconomic and environmental aspirations of local communities, conservation organizations, and local authorities.

Research results

Wetlands in Ghana have been exploited for their environmental resources, which vary according to the specific environment. Traditionally, they have been exploited primarily for subsistence, usually in conjunction with the resources of the adjoining nonwetlands. Commercial exploitation is limited and is most important in the coastal wetlands, where salt-winning may be a major activity. Modern resource exploitation combines both commercial and subsistence systems. To understand the economic, social, and environmental issues, it is important to understand how wetlands are currently being used in Ghana: traditional uses of coastal wetlands, the traditional uses of noncoastal wetlands, and the modern uses of both coastal and noncoastal wetlands.

Traditional uses of coastal wetlands

The coastal wetlands are traditionally exploited for agriculture, fishing, salt-winning, and natural products such as palm leaves.

Agriculture is relatively unimportant in the coastal wetlands. Crops are grown only on elevated sites or high ridges. Staple crops, including cassava and maize, are grown using the bush-fallow system. Vegetables are also grown, including onions, okra, peppers, tomatoes, water melons, and beans. Sugar cane is locally important in the estuarine wetlands, where there is freshwater. Coconuts are widely tended along the coast. Coastal wetlands also provide ample grazing materials for livestock. Chicken, ducks, pigs, cattle, goats, and sheep are common in all the settlements there.

Fishing is a common and important activity in all the coastal wetlands, especially in lagoons, but also in the brackish-water marshes in depressions. The equipment used includes cast nets, draw nets, and traps. The most important lagoon fish, the tilapia (*Saratherodon melanotheron*), accounts for about 98% of the catch. Fish are locally important, not only for the cash they bring in, but also for the protein they provide in the otherwise protein-deficient diets of the wetlands communities.

Salt-winning is a term used to describe a process of collecting salt from the lagoon floors during the dry season. High evaporation leads to the precipitation of salt, which traditionally is then freely collected by both residents and nonresidents. For local communities, salt-winning may be the dominant way to generate income, along with sea fishing. Other products collected from the coastal wetlands include fuelwood (for cooking and for smoking fish) and grasses and sedges (for thatching and mat making). Collecting shells is locally important for the production of quicklime and chicken-feed aggregates.

Although the coastal wetlands have the potential for tourism, it has developed little outside the beach environments.

Traditional uses of noncoastal wetlands

The noncoastal wetlands, like the coastal ones, are exploited for agriculture, fishing, and natural products. Crustaceans are also collected

Agriculture is important in large portions of the lands bordering the Volta. In the savanna-based wetlands, staple crops, such as maize, millet, and rice, are grown. Vegetables are cultivated on a commercial scale in the more accessible inland wetlands, such as in the southern reaches of the Afram Plains. In the forest region, the lands adjoining the permanent marshes are cultivated using the bush-fallow system. Important crops include plantain, bananas, rice, and maize. Root crops are of minimal importance. Sugar cane is an important crop, for example, between Nsawam and Suhum. The wide range of vegetables includes tomatoes, okra, and peppers. Rice is grown in rotation in the small valley bottoms in both the savanna and the forest areas. The Western Region is a major area for rice cultivated in this way. Since the 1970s, mechanical land preparation, fertilizer, and other chemical inputs have been used to grow large acreages of rice in the valley-bottom wetlands in the Northern Region. These areas are developed downstream of dams.

Dry-season grazing of cattle, sheep, or goats is common in the wetlands of the northern savannas. Grasses and shrubs persist much longer in the bottom wetlands, after clearing, harvesting, and burning have removed most vegetation from the dry uplands. Also, these wetlands, particularly the streams, provide the animals with plenty of drinking water.

Freshwater fishing in the noncoastal wetlands and streams is as important as it is in the coastal wetlands. In the forest regions, crabs are caught, in addition to the fish. Large freshwater crabs are caught in most wetlands, such as between Kibi and Nkawkaw. Freshwater shrimp and snails may be collected, although these two sources of protein seem to have been eliminated in many areas by the increasing degradation and pollution of habitats.

Collecting natural products is particularly important in the forest wetlands. Principally, *Raphia* (palm) is used as material for construction, roofing, and abrasives. The palm is also tapped for wine. Leaves from the higher ground surrounding the wetlands may be collected for medicines and wrappers.

Modern uses of both coastal and noncoastal wetlands

In the latter half of this century, a number of Ghana's wetlands have been managed for irrigation and large-scale salt-winning. Most recently, wetlands have been managed as Ramsar sites for the conservation of biodiversity.

Valley-bottom irrigation below dams was part of the Land Planning program in the northern savannas in the 1950s. Land Planning was a conservation program to rehabilitate and protect land and water resources in densely settled lands (Lynn 1945). The irrigated valley bottoms were usually small. During the 1960s, large dam projects inundated valley-bottom wetlands upstream and drained others downstream. Flood-plain wetlands downstream were converted into large-scale irrigation areas, usually for the commercial production of rice, sugar cane, and vegetables. The upstream areas became reservoirs that are used for fishing and provide water for domestic, industrial, and farm use and for hydroelectric power. These large schemes are located, by and large, in the dry-coastal and northern-savanna environments.

Lake Volta, created in 1964, is the largest such scheme: it covers 8 480 km^2 and has 5 200 km of shoreline. Other large dams include the Vea and Tono, in the upper East Region, and the Okyereko and Komenda, in the Central Region. Owing to the difficulties of managing the large dams, attention has been directed to developing the smaller valley bottoms, which often contain small wetlands.

The Irrigation Development Authority, Ministry of Agriculture, has collaborated with the Agricultural Development Corporation of Korea to earmark pilot irrigation schemes in some areas of the valley bottoms in the forest zone. Their feasibility studies have included topographic, soil, agronomic, and socioeconomic surveys, and they have completed systems layouts for a few schemes, such as at Kikam, Oda-Bekwai, and Akim Abodom. The Anum Valley irrigation project, near Konongo in the Ashanti Region, for example, involves the reclamation of 200 ha of valley-bottom land for paddy rice cultivation; a system of water-restoring dams is involved. Aquacultural development may also be part of the project (IDA 1990).

Wetlands conservation

Wetlands conservation in Ghana focuses mainly on the conservation of water resources and biodiversity. Water conservation involves damming rivers for a more effective use of water for agricultural, industrial, and domestic purposes.

Wetlands management traditionally included a conservation component, reinforced by custom, religious taboos, and totem systems. During sacred days or periods, fishing was prohibited in streams and lagoons. Certain plants and animals were protected. Despite the effectiveness of these traditional conservation practices, modern wetlands conservation has not been based on them.

The Save the Seashore Birds Project of Ghana, initiated in 1986 and sponsored by the Royal Society for the Protection of Birds, has identified 13 key coastal wetlands that provide feeding, roosting, and nesting sites for thousands of migratory and resident seashore birds. EPC, under the biodiversity component of the Ghana Environmental Resources Management Project (GERMP), has prepared management plans for the conservation of five of the coastal wetlands as Ramsar sites, that is, internationally important wetlands conserved through multiple-use management. The five sites are Songor, Sakumo, Densu Salt Pans, Anlo-Keta, and Muni. Four others may also be similarly protected: the Keta, Korle, and Amazuri wetlands and the Elmina salt pans.

The general objectives of these Ramsar sites are

+ To maintain and enhance the value of the wetlands as wildlife habitat and to integrate wildlife conservation into the current human use of the wetlands;

+ To enhance the benefits derived from coastal wetlands and to improve the quality of life for the local communities in the vicinity of the wetlands; and

+ To control, monitor, and coordinate activities affecting the coastal zone (for example, human settlements, industrial development, salt production, agriculture, fisheries, and recreation) so that the integrity of the coastal environment and the sustainability of wetlands resources are maintained.

The noncoastal wetlands are not generally protected in these ways, although some may be if they are within existing forest reserves.

Settlement development

Some large wetlands complexes have been settled (as is the case with the Songor and Anlo-Keta lagoon wetlands). Management of these areas has become complex and challenging because of the threat of environmental degradation. Lagoons and lakes may be polluted; fish stocks and other wildlife, depleted; and shrubs and trees, such as mangroves, cut for firewood and construction materials. Where wetlands consist of zones that grade from wet and dry areas to permanently shallow water, settlements are located in surrounding areas. In this case, the availability of nonwetlands resources may reduce pressure on the wetland. Exploitation of wildlife is limited to one or two useful species, and the difficulty of the environment keeps the soils from being used. This is the case in the valley-bottom wetlands of the forests. The problems created by the simulium fly in the valleys in the northern savannas have restricted settlement in and exploitation of the valley-bottom wetlands in those areas (Hunter 1967).

General observations

Ghana's wetlands vary in size and water regimes. Generally, data and information on the wetlands environments and their use are lacking (Agyepong 1993). Traditionally, wetlands have been regarded as important by some people and of minor significance by others. Likewise, traditional uses of wetlands vary from dominant to minor and supplementary. However, the use values and conservation values of the wetlands are recognized in programs for agriculture, aquaculture, salt development, and biodiversity conservation. The low topography of the wetlands makes them vulnerable to pollution and degradation in areas where settlement and agriculture are increasing.

Analysis

Wetlands resources need to be sustained for their unique, economic, environmental, and social values. The study suggests that the wetlands in Ghana provide a broad range of social, economic, and environmental benefits:

+ Offering opportunities for sustainable socioeconomic development and improvement in the quality of life of most people in Ghana;

+ Serving important ecological functions, including provision of habitat;

+ Promoting the socioeconomic welfare of people living nearby;

+ Giving incentives for the development of settlements and communities that thrive on wetlands resources;

+ Creating jobs and promoting strong sectoral linkages; and

+ Providing avenues for retail trade, which benefits almost all urban areas in the country.

Nevertheless, the current degradation of wetlands ecosystems in Ghana is significant. The study found that

+ All types of wetlands in Ghana have suffered to a greater or lesser extent from some form of degradation resulting from externalities;

+ Industrial and agricultural activities have combined with sewer effluent to degrade the chemical and biological quality of wetlands;

+ Most wetlands close to urban areas have been degraded through anthropogenic activities; and

+ Wetlands degradation in Ghana results from land-use conflicts, information and market failures, and inefficient government policies.

It is therefore important to conserve wetlands, not just as units in the landscape, but as a resource base for the socioeconomic development of the nation. The following are observations of current wetlands-conservation practices in Ghana:

+ Wetlands conservation in Ghana is mainly based on traditional beliefs, taboos, and norms, which are undocumented but handed down orally from one generation to the next;

+ Because of social and economic transformation, traditional conservation practices have been unable to halt the long-term degradation of wetlands resources; and

+ Recent conservation efforts have been aimed at designating wetlands as Ramsar sites.

Socioeconomic-infrastructure development

Social services and infrastructure of the wetlands in Ghana may be characterized as follows:

+ Wetlands communities lack basic social services and technical infrastructure, and accessibility to these services is also minimal;

+ Very few wetlands communities have access to education, clean water, health, and electricity; and

+ Most wetlands communities suffer from low per capita income, poverty, and ignorance.

Toward sustainable wetlands management

The wetlands research project came to a number of broad-ranging conclusions about the development of sustainable wetlands management in Ghana:

+ Wetlands management should aim to sustain wetlands resources for their unique economic, environmental, and social values;

+ Sustainable wetlands management would be not only helpful in maintaining the balance between tangible and intangible benefits but also essential for the very welfare and survival of most Ghanaians; and

+ Sustainable wetlands management would provide habitat for numerous flora and fauna and generally maintain biodiversity.

Implications for policy and institutional frameworks

Wetlands provide comprehensive environmental and habitat protection. It is thus important to give priority to restoring and maintaining the wetlands' natural vegetation and conserving their natural quality. Sound management of the wetlands nationwide can be promoted in either of two ways: by explicitly including wetlands management in existing environmental policies or by creating a separate wetlands-management policy.

Developing a sound base for wetlands management in Ghana involves two key issues: the management strategy that should be embarked on; and essential elements that need to be incorporated into the management plan for the wetlands.

Wetlands-management strategy

GERMP has established a project component for wetlands management. Under this program, a review was conducted of the functions of Ghana's wetlands. This preparatory project showed that most of the stress occurs in wetlands of international significance for flora and fauna (especially wildfowl). Consequently, certain wetlands have become a high priority in Ghana's national conservation policies. Because these are designated as Ramsar sites, no specific policies are needed to protect their functions, as these are taken care of under the general wildlife-conservation policies.

Each wetland is unique, both in terms of economic factors (because they usually have no alternative location) and in terms of environmental factors. A good wetlands policy must be designed and agreed to by the various actors and interest groups responsible for their development and should integrate environmental, social, and economic factors.

Designing and implementing a wetlands policy

Because the ministries of Food and Agriculture, Land and Forestry, Environment, and Science and Technology have no clear definition of wetlands, the National Development Planning Commission should be responsible for a separate wetlands-development policy.

The current Wildlife and Protected Areas Policy emphasizes wildlife resources, with little regard for the wetlands as a whole. This policy seeks to establish and maintain the sustainability of the marine and other protected areas, with the eventual aim of protecting a minimum of 10% of Ghana's land surface and managing all protected areas according to detailed management plans.

Policy limitations

The Ghana Forests and Wildlife Policy covers only those wetlands designated Ramsar sites. Its intention is to ensure that viable populations of all indigenous wildlife species, including passage migrants, are adequately conserved and that rare, endangered species of high conservation interest are especially protected. Having only this policy endangers the other wetlands of local importance, which is one of the most important findings of this study.

Also, the agricultural sector has failed to develop any mechanism to systematically differentiate between projects affecting wetlands and those affecting other water resources and drainage. Lack of definite policies regarding the use of wetlands resources puts wetlands in Ghana at a risk. The need for a wetlands-management policy for Ghana is strongly apparent.

Wetlands-management plans should include guidelines for designing wetlands policies, improving the production base of the communities, and providing well-integrated environmental programs.

Impacts of the research on government policy in Ghana

This research has had impacts on policy formulation at three levels of government, that is, at the national, regional, and local levels. At the national level, the research drew a lot of attention to the identification and mapping of noncoastal wetlands. The inland wetlands had never been identified for conservation efforts. The government of Ghana now has a strong interest in designing programs for the conservation of inland wetlands.

At the regional and district levels, the material in the wetlands-degradation report persuaded municipal authorities to call for severe restrictions on industries polluting wetlands, as well as for the control of domestic-waste discharge into such waters.

At the grass-roots level, there is now an interest in conserving wetlands of local importance as complements to the wetlands of international importance under the Ramsar Convention. Various local authorities have taken up the challenge, such as those in Esuekyir, a small fishing community in the Gomoa Effutu Ewutu District of Southern Ghana.

Guidelines for a wetlands-management policy

The overall national-development strategy in Ghana should recognize the environmental, social, and economic importance of the wetlands. Therefore, wetlands policy for Ghana should be part of a general environmental policy, allowing wetlands-management objectives to penetrate decision-making within government administration. Wetlands issues should have a relatively high priority on the environmental agenda. Dialogue and coordination between public bodies in charge of the environment and their counterparts in the economic sectors should be promoted, as well as mobilization and social welfare. This will enhance the sharing of experience and priorities and help in developing integrated environmental, social, and economic policies.

For effective management of the wetlands, environmental policies should be integrated into current development policies. This is necessary to ensure development of technically and economically feasible, socially acceptable, and environmentally appropriate uses for the wetlands as an alternative to either purely economic or purely protectionist approaches. To formulate a national wetlands-management policy as an integral part of the general-development policy, the government should

+ Create or assign an agency to be responsible for overseeing sustainable management of the wetlands and to coordinate, control, and harmonize policies that affect the status of the wetlands; and

+ Strengthen the requirements for environmental-impact assessments so that assessments will be more critical of projects likely to affect wetlands conservation.

Toward a system for integrating economic, social, and environmental policies in Ghana

Experience from this wetlands-management study in Ghana shows that the design and implementation of integrated policy should be a continuous process that provides the prospect of improved administration and effective planning and implementation of integrated-development policies. Procedurally, a new system, in accordance with government policies, should be integrated, decentralized, participatory, problem solving, and continuous.

Integrated

The system should be designed to analyze each development issue from the political, social, economic, environmental, and spatial perspectives as a single, integrated task. The result of such analysis should then be synthesized in comprehensive, cross-sectoral, and mutually reinforcing development programs.

Decentralized

The development process — from goal setting to the preparation of medium-term strategies, implementation programs, annual action programs, and budgets — should be decentralized and integrated horizontally and vertically. This integrated development process should be a collaborative exercise, with functional agencies, sectoral ministries, regions, districts, communities, and the private sector playing clearly identified roles at all levels.

Participatory

Communities and the private sector have important roles to play in developing integrated policy, as such policy must relate to people's aspirations and priorities. Community participation needs to be made central to the development process and should never be taken for granted. This brings a sense of reality into development decision-making and harnesses local expertise.

Problem solving

At the local level, information is expensive to collect and process. The integrated system should target identified priorities, problems, and needs. Planners have often devoted considerable resources to answering questions no one was asking. To be effective, a good policy must solve the problems of concerned communities.

Continuous

Socioeconomic systems are in a continuous, dynamic process of change — through space and time — and the development process must respond to this. Data collection, updating, analysis, and synthesis and policy implementation, monitoring, and evaluation must proceed concurrently. This will enable policy-makers to maintain policy equilibrium with adjustments for changing circumstances. Long- and medium-term policies and all strategies and programs must therefore be subject to periodic review and revision.

Operationalizing the system

In Ghana, a new integrated planning system commenced with the National Development Goal Setting Exercise. The National Development Planning Commission studied policy statements in government reports, ministerial statements, and the press, which were translated into technical-development goals and objectives. The secretariat of the Commission collated these responses and reformulated them as the Hierarchy of Long Term National Development Goals and Objectives. The Cross Sectoral Planning Groups for Human Development, Economic Development, Spatial Organisation and Environment used this as the control in preparing four more-detailed hierarchies of goals and objectives, according to subject areas. These five hierarchies were then used to prepare the

National Development Policy Framework (the Vision 20/20 project), a 25-year perspective for the transformation of society and the economy over a generation (NDPC 1991, 1995). This is expected to be a well-integrated policy framework, addressing social, economic, political, and environmental concerns.

The governance of integrated policy at the local level

Political administration in Ghana is carried out at the national, regional, and district levels. The national government designs the policy framework, whereas the regions play a role in monitoring and coordinating. Because of their proximity to the people, the district assemblies play the critical role of problem identification and policy implementation, a role that demands considerable ingenuity and innovation. According to *Local Government Act 462* of 1993, the most important activities of the district assemblies include

+ Acting as an agent of change to transform the economy;

+ Identifying and promoting development opportunities;

+ Creating an enabling environment to help the private sector contribute to economic growth and social well-being;

+ Mobilizing human and physical resources;

+ Coordinating all development activities in the district;

+ Protecting vulnerable groups, particularly women and children;

+ Disseminating information on development issues; and

+ Facilitating public participation, as an integral part of the development process.

These activities each directly or indirectly contribute to the primary task of the local authorities, creating an enabling environment. The effective discharge of these responsibilities requires a high level of commitment, competence, and efficiency and significant changes in the attitude of public servants.

Local authorities in Ghana indicated that the conditions for implementing integrated policies should include

+ Continuous commitment at the national level, including from Parliament, to integrated programing and to devolution of power at all levels;

+ Effective capacity-building within a transformed, restructured public administration;

+ An enabling environment for the private sector, districts, and communities;

+ Discipline and an effective incentive system for sustainable development; and

+ Education programs for people to learn about the meaning of integrated sustainable development and the potential for ordinary citizens
to change economic and social conditions for the better, as well as for
the future generations.

Institutional arrangements for integrated policy and planning in Ghana

Since Ghana achieved independence in 1957, a greater awareness of the effects
that institutional structure and decision-making frameworks have on design and
implementation of integrated policies has emerged. Both Trevallin (1994) and
the Action Program for the Rehabilitation of Wetlands in Ghana (FOE–Ghana
1994) suggested some key concepts to guide institutional arrangements for integrated policy and planning:

+ Close association of planning agencies and the political decision-making
bodies;

+ Centrality of integrated-planning activity at all levels of government;

+ A need for a single agency that is responsible for coordinating integrated planning and policy at each appropriate level;

+ Combined physical, economic, social, environmental, and cultural
planning within this agency;

+ A need for effective implementation, with feedback in the process;

+ Clearly defined responsibilities for integrated planning, with roles,
levels, and objectives defined in legislation; and

+ Society's ultimate responsibility for choosing strategies.

Formation of a national-level integrated-policy agency

A single agency should be responsible for the environmental, social, and economic policies of the country at the national level. This agency should have
expertise in economics, sociology, and environmental policy and planning for
national, regional, urban, and rural development.

This national-level organization should comprise

+ A process group for formulating national planning strategies, perspectives, and guidelines for integrated development;

+ A specialist group for urban development;

+ A specialist group for rural development; and

+ Sectoral groups.

These groups must have a balanced complement of both physical scientists and socioeconomic analysts. Without this, integrated planning expertise cannot be brought to bear on the development-capital budgeting processes in an effective way, which is essential. This complement will also ensure that these groups understand how the social, environmental, spatial, and economic aspects of development problems interact and strengthen each other.

Recommendations for functions, roles, and outputs of the integrated-policy agency include the following:

+ The organization should adopt a problem-defining and problem-solving approach to national development problems. Initially, these problems will stem from the physical and spatial constraints on achieving national objectives.

+ The organization should maintain a continuously updated data bank pertaining to national development. This should be in a format usable by planning agencies at every level and be available as well to the sectoral development units and other appropriate government and quasi-government bodies. It should be progressively developed by a structured, continuous, and monitored study program.

The output of the agency's activities should include

+ Broad physical, spatial, social, and economic strategies for development at the national level;

+ Definition of subject themes and geographical areas for integrated priority and programed action at national, regional, and district levels;

+ Monitoring of the appropriateness of strategies and programs during and after implementation, with feedback into an ongoing planning process; and

+ National perspectives on resource consumption, urbanization, settlement growth, and population distribution.

Strategies should be continuously updated. The national strategy should comprise the equivalent to current long-term economic perspectives and deal with long-range hierarchical, spatial, physical, and economic characteristics of resource development, settlements, and the environment. These should be filtered through to the grass-roots level through the regional bodies.

It must be stressed that the regional strategy for development should not be a repetition of the national framework. Regional-level planning for social and economic development ensures the practical execution of the nation's objectives within local culture and involves harnessing the people to national endeavour.

Recommendations for the local planning organization include the following:

+ It should include a process group concerned with the interpretation of national and regional development policies for implementation at the local level. This group should also be concerned with the development of strategies and priorities in response to local objectives.

✦ It should have a development management and implementation group to prepare detailed, costed action programs within the overall strategic framework. This group should also continuously monitor and evaluate the effectiveness of action programs to fulfil society's development goals. Information should be fed back into the ongoing planning process.

✦ Local planning agencies should be separate bodies within the local authorities, but the planning process must be integrated with the activities of the specialist departments. A mechanism must be devised to make this essential integration of work mandatory and continuous.

✦ A mechanism must be devised to enable the process group to continuously study the interrelationships of the local, regional, and national physical, spatial, and economic components of development.

Conclusions

An important conclusion of various studies on the relationship between environment and development is that environmental, social, and economic considerations need to be integrated in development activities.

Coastal wetlands have been a focus of policy analysis in recent times in Ghana. Wetlands in the interior savanna zone have not received much attention to date. Traditionally, these areas have been exploited for subsistence, usually in conjunction with the resources of the adjoining nonwetlands. More recently, developed exploitation practices have combined commercial and subsistence systems to include commercial agriculture, salt extraction, and settlement development.

The continual changes in government structure and roles of integrated-planning agencies in Ghana have created inconsistencies in policy design and implementation. Problems include an overemphasis on sectoral policies, with little or no integration; monolithic and overcentralized planning for development; and lack of feedback from the local level. In formulating broad policies, policymakers have given very little attention to local-level development that meets the aspirations of local people. The result is that communities feel alienated from development priorities, and local people end up implementing sector policies that do not "belong" to them. A meaningful integrated policy would carefully take into account the interplay of economic, social, environmental, and political factors.

A policy-related decision-making structure to effectively address this concern should be

+ Integrated (that is, each issue should be analyzed from political, social, economic, environmental, and spatial perspectives as a single, integrated task);

+ Decentralized, to promote collaborative efforts, with stakeholders playing a meaningful role;

+ Participatory, to make development policy responsive to the aspirations and priorities of people;

+ Problem solving, or targeted to the solution of identified problems and needs; and

+ Continuous, with ongoing data collection, updating, analysis, synthesis, monitoring, and evaluation and a well-defined feedback mechanism.

The greatest challenge will be establishing the institutional framework for implementing such integrated policies. This will require effective and proactive capacity-building to ensure an enabling environment for the private sector, district authorities, and local communities and an effective incentive system to create a sustainable and equitable society. In the case of wetlands management, such a framework should be capable of ensuring that international, national, and local priorities are met in conjunction with the broader goal of the long-term sustainability of wetlands resources.

Chapter 7

Supplying Clean Water to the Citizens of Nairobi

P.M. Syagga

◇ ✦ ◇

Introduction

Integrating environmental, social, and economic concerns in planning for sustainable development has received considerable attention in recent years. Chapter 8 of Agenda 21 (UNCED 1992), signed at the Earth Summit, declares that environmental protection is at the centre of the sustainable-development process and cannot be considered in isolation from social and economic development. The increased interest in integrated policy-making stems from a recognition that the separation of economic, social, and environmental factors in decision-making has not been conducive to sustainable development. However, much remains to be learned on how best to adjust and reshape decision-making processes to fully integrate environment and development activities. Research is needed to examine the conditions that facilitate and promote integrated-policy development. In addition, because research often represents a first step in bringing about a change in the way decisions are made, research processes need to be developed that in themselves provide a basis for building integrated-policy processes. The purpose of integrated-policy research is to inform policy-making and promote the design and implementation of policies that reflect the interdependence of social, economic, and environmental systems. This chapter presents some observations on integrated-policy research from an African perspective and illustrates these with a case study of a research project to evaluate the Third Nairobi Water Supply Project in Nairobi, Kenya. This is followed by an analysis of the case, highlighting both the strengths and weaknesses of the research project, with particular emphasis on links between research and policy.

Integrated-policy research and planning in East Africa

The overall objective of integrated policy is to promote sustainable development in the context of policies that integrate efficient use of environmental resources, reduction of poverty, and strengthening of human and institutional capacity for development. In Africa, environmental degradation and poverty combine to create a vicious cycle, with poverty a key factor in producing environmental degradation and vice versa. This cycle can be narrowed and eventually broken if sustainable development, as an environmental-management system, becomes the norm.

Despite this gloomy picture of economic and ecological crisis in Africa, many countries still take a purely conservationist approach to environmental management, rather than one that focuses on the broader context of sustainability and simultaneously addresses both human and resource sustainability. In much of Africa, the linkages between environment and development continue to be overlooked, as evidenced by sectoral-development programs without a structural environmental perspective. A methodology linking environment with economics or social policy remains unclear and elusive, a result of the lack of institutional development, appropriate techniques, and an enabling political climate.

Integrated-policy concerns first appeared in the development literature in Africa during the 1960s. In essence, the literature called for increased attention to the interrelationship between individual aspects of rural development and for the implementation of packages of mutually reinforcing activities in the rural areas. Various attempts to undertake a set of interrelated activities simultaneously were made in projects such as the Gezirah irrigation project in Sudan and the Accelerated Rural Development Programme in Botswana (Sterkenburg 1987). These and many other projects in East Africa reflected a basic-needs approach to development, employing measures directed at generally improving the living conditions in the rural areas, without identifying a target population.

In more recent times, the concept of integrated development has been applied in a variety of situations, in both rural and urban areas, but with emphasis on target groups, particularly the poor. Multilateral and bilateral agencies, national governments, and nongovernmental organizations (NGOs) are sponsoring programs that incorporate integrated activities as a development package, which, for lack of a better term, researchers in the region have named the "package approach." Housing projects in East African cities, such as Dar-es-Salaam, Kampala, Lusaka, and Nairobi, have targeted low-income groups, used site and service schemes and squatter-upgrading projects, and incorporated activities to improve housing conditions and activities to promote income generation. This represents a move away from the sectoral approach that, for instance, considered housing development only as shelter provision.

Accordingly, *integrated policy*, as used in East Africa, refers to a development policy that addresses a package of issues and aims at improving the living conditions of a target group. Solutions to issues are to be implemented simultaneously as an integrated set of activities. As emphasized at the Earth Summit, human beings are at the centre of concerns for sustainable development. If sustainable development is to be achieved, environmental protection must be considered an integral part of the development process — it cannot be considered in isolation. Because of the delicate ecological base of most of the countries in Africa, environmental concerns are achieving political prominence, particularly within regional bodies such as the Inter-Governmental Authority on Drought and Development. Although the conservation message has been given political legitimacy in some countries, available evidence suggests that the capacity of the existing institutions to develop and implement sustainable-conservation programs is limited (Kiriro and Juma 1989). The available knowledge on natural-resource management and economic and social policy is still limited, and more systematic efforts are needed to collect and analyze information to enable key policymakers and practitioners to identify feasible policy options.

Environmental awareness in urban areas is increasing, particularly in light of deteriorating urban infrastructure and the problems of waste management. A number of cities now implement programs that incorporate environmental concerns, such as the Sustainable Cities project in Dar-es-Salaam and the Green Towns project in five secondary towns in Kenya. Still lacking are legally enforceable standards or codes of practice on environmental issues to facilitate monitoring.

Efforts have been made to include approaches to development similar to those of Agenda 21 in the project packages. The objectives of Agenda 21 can only be reached, however, if the region fully appreciates the significance of environmental concerns in social- and economic-development planning.

Key features of the package approach

The package approach to development policy in East Africa has included various aspects of integration in the planning and implementation processes in specific projects. The following are some features of integrated-policy research and planning in East Africa;

Inter- and intrasectoral integration

Inter- and intrasectoral integration is based on specific functions and resource positions and incorporates inter- and multidisciplinary tenets of integrated-policy research. For example, for project implementation in the present case study, an Inter-Ministerial Task Force was formed, incorporating key ministries, particularly those with portfolios for water, environment, agriculture, and provincial administration.

Target-group participation

Target groups participate in the planning and decision-making process, particularly in projects targeting poor communities. The degree of participation has depended on the type of activity, as well as the implementing agency. Projects executed by international and national NGOs have sought to obtain full partic-ipation of the target groups, from inception to completion, by operating through community-based organizations. Projects implemented by government agencies, on the other hand, use civil servants as agents of change; participation of target groups is limited and primarily at the implementation stage. Clearly, there is vertical integration in the package approach, although the degree of inte-gration is variable. In the present case study, an Environmental Action Programme committee was formed as a subcommittee of the Inter-Ministerial Task Force and included all government-department heads, NGOs, and repre-sentatives of the affected population in the project area. The committee was to oversee project implementation on the ground. However, in terms of integrated-policy research, vertical integration is seen to be achieved only when the benefi-ciaries fully participate, from project planning right through to implementation. The beneficiaries might include community groups and other stakeholders, such as policymakers and advisers. Thus, the integration process in this local context has not yet fully realized its potential.

Administrative, or horizontal, integration

Administrative, or horizontal, integration is based on multi- and interdiscipli-nary activities within the package. These activities may or may not involve envi-ronmental concerns, as no deliberate efforts are made to integrate these concerns with economic or social issues. Often, environmental concerns come into con-sideration with issues like population and agriculture or in studies on human settlements. Poverty studies, particularly in the urban context, seem to be con-cerned with income generation and health, rather than with environmental degradation per se.

The extent of horizontal integration of leadership, the working arrange-ments, the inclusion of various disciplinary perspectives, and the type of research tools depend on how the research project was initiated, who prepared the brief, and what type of award systems are in place to promote involvement and inte-gration. Often, the disciplinary perspective prevailing at the initiation of the project is the perspective that is assumed by those who take up the role of team leadership as principal researchers, and that influences decisions about the disci-plinary perspectives of people invited to participate.

In this scenario, researchers from each discipline are asked to prepare a proposal on how the activity or task should be carried out, as well as estimating the cost and time involved. Each researcher is expected to execute the research independently and present a draft for the team leader to incorporate into the whole report. Invariably, the final research document is structured in sections, according to the inputs of the researchers' various disciplines. For instance, in this case study, the initial feasibility study was carried out by an engineering con-sulting firm, with an engineer as the team leader. Researchers from other disci-plines were asked for their inputs; however, the engineering concerns about the dam construction were most prominent. The project's involvement of the affected population was minimal.

In sum, integrated research in East Africa still has some way to go and may require deliberate efforts to realize its potential. At the outset, both researchers and policymakers may see integrated-policy research as synonymous with participatory research.

Participatory research

In participatory research, the researcher has a subjective commitment to the target community and to the betterment of human conditions. Thus, in theory, the target population is given a chance to express its views on relevant solutions to its problems, and more research time is spent in and with the community so that community members can participate in the identification of research questions, the selection of methods of data collection, and the analysis. Various forms of knowledge, including valid popular knowledge stemming from people's sociocultural heritage and practical experience, are accepted. However, in practice, difficulties arise in choosing community representatives. The complex reality of power relations in the community acts to favour individuals having different levels of participation, depending on factors such as age, gender, socioeconomic status, and family background. In many cases, the participating group may not be representative of the whole community. In participatory research, as well as in integrated-policy research, greater benefit derives from widening the spectrum and increasing the number of people who participate.

This need is often addressed by selective participation strategies, recognizing that existing local forms of decision-making that provide for popular representation may not reflect the values and norms for participation held by outsiders.[1] Although participatory research is also action oriented, it has a vertical-integration methodology and therefore falls short of horizontal-integration requirements for integrated-policy research. Integrated-policy research is not only participatory but also multi- and interdisciplinary in approach and gives special consideration to the interdependence of social, economic, and environmental concerns. This is yet to be fully realized in East Africa, and deliberate actions of researchers and policymakers will be needed to meet the demands of integrated-policy research.

Methodology of integrated-policy research

Policies and programs for sustainable development can be successful only if the people affected by these policies and programs fully participate in their development and implementation. As a rule, the majority of decision-makers at the technical level in most governments and development agencies in East Africa are social scientists; only a few are natural scientists, and they have failed to apply

---

[1] Syagga, P.M. 1994. Integrating environmental, social and economic policies (INTESEP) in urban planning and development research in eastern and southern Africa. *In* IDRC highlights report. INTESEP workshop, Abidjan, Côte d'Ivoire. International Development Research Centre, Ottawa, ON, Canada. Unpublished manuscript. Annexe 1. pp. 1–23.

the available scientific knowledge to human situations. Under the influence of politicians, a tendency has been to go for easy options with short-term gains, without due regard to long-term sustainability.

The prevailing assumption among many policymakers is that politicians at local and national levels represent the community and that involving these representatives ensures the full participation of the community. Researchers have also thought that politicians hold the goodwill of the community and that the politicians' involvement secures community acceptance. Consequently, numerous intervention programs in development have failed. The need to ensure community acceptance for the sustainability of interventions has led many to look at the value of participatory methods to help match beneficiary needs and realities with development initiatives and thereby contribute to the greater local control and sustainability of these efforts (Baldwin and Cervinskas 1991).

This is particularly relevant to integrated-policy research. In terms of the research–policy nexus, it is essential to use the tools of participatory research to promote real dialogue, horizontally between researchers and vertically between policymakers and beneficiaries. The fact that we live in a dynamic society forces us to revise the traditional perspectives and methodological tools we use to investigate development problems. A new approach is needed to define research problems, gather and analyze data, and report information to advise policy. New types of arrangements within and between institutions, beneficiaries, and decision-makers may also be needed to effectively use such research methods. The underlying issues are bringing the different disciplines to adopt a common approach to development issues and getting policymakers to appreciate the contribution of the community in the research process.

Planning

Integrated-policy research involves greater collaboration, cooperation, and communication among the research team, policymakers, and beneficiaries. It thus needs both horizontal and vertical integration. Accordingly, integrated-research preparation (problem definition and team selection) should be a participatory process, involving multistakeholder partnerships, so that a shared understanding of the problem develops. An effective team leader is essential to the success of the project. Because integrated research addresses social, economic, and environmental concerns together, the different disciplines must adopt a common approach to the research design and implementation at the planning stage.

Data collection and analysis

To ensure appropriate integrated-policy research, the team leader should balance the input from each researcher but remain focused on the problem and not be distracted by competing disciplinary assumptions. Analytical tools and concepts used by one discipline should be transferable to and readily understood by researchers from another discipline. Collaboration, cooperation, and communication in data collection and analysis should be ongoing among the researchers, beneficiaries, and policymakers. The researchers from different disciplines should agree on a common set of parameters, a common level of significance for accepting the research results, and a common format for presenting data. All the other actors in the research process should be involved in reaching this common understanding.

Presentation of findings

Given that integrated-policy research aims at informing policy-making and promoting interdependence of policy concerns, the presentation of research results should focus on achieving cooperative agreements for change. The approach should build trust and confidence among partners as a basis for a common agenda, common priorities, and common actions. The presentation should also tap and make more effective use of local and indigenous knowledge and skills and be a capacity-building process. A synthesis of research should be presented in a format amenable to existing policy mechanisms and should be followed up with a workshop with all stakeholders and a proposal for a plan of action. All this will help promote an effective research–policy nexus.

The practical role of research in integrated policies

This section examines a specific effort to promote integrated research in the design stage of a research project. The shortcomings of this project provide a number of lessons for the design and implementation of future integrated-policy research.

The Third Nairobi Water Supply Project evaluation

Background

Since Nairobi became the administrative capital of Kenya in 1900 a number of water-supply projects have been undertaken, but until the Third Nairobi Water Supply Project, none involved an environmental-impact assessment (EIA) before implementation, particularly with respect to residual impacts of the water abstraction and the sludge-waste disposal into the rivers. Also absent from these projects were assessments of their socioeconomic impacts.

As the population of Nairobi was growing at 6% per year and would treble by 2010, Nairobi City Council (NCC) embarked in 1988 on the Third Nairobi Water Supply Project, expected to yield 460 000 m^2 of water per day (Syagga and Olima 1996). The estimated cost of the project was 30.5 million US dollars, and the project was cofinanced by the Overseas Economic Co-operation Fund of Japan, the African Development Bank, the European Investment Bank, NCC, and the Government of Kenya. The three funding agencies were represented by the World Bank, and their funding was channeled through the World Bank. When completed, the project was expected to displace 500 households and flood 350 ha of land used for small-scale tea farming. In addition, the water was to be pumped to Nairobi in 4-m-diameter pipes, constructed on 24-m-wide wayleaves acquired through private parcels of land, over a distance of some 60 km. Although the project site was in a rural district, and hence the affected people were in a rural settlement, the ultimate beneficiaries were to be the residents of the City of Nairobi.

Recognizing this project would have both positive and negative impacts on the population and the environment, NCC commissioned a number of studies to prepare plans for an environmental-action and -monitoring program (Howard Humpreys [Kenya] Ltd and Environmental Resources 1988; Syagga Associates 1988; Acquasystems Consultants 1989). These plans were to ensure that few negative socioeconomic and environmental impacts resulted from the implementation of the project. For instance, NCC had to initiate and facilitate investigation of any negative effects on the environment, including soil erosion and pollution, from construction and operation of the dam and its aqueducts. NCC was also required to pay prompt and adequate compensation to the displaced families and work out resettlement programs to minimize personal suffering, as provided under section 75 of Kenya's Constitution and the *Land Acquisition Act*, chapter 295 of the *Laws of Kenya* (GOK 1968). However, these commissioned studies were carried out independently by the three firms, and the implementation of the plans seems to have been considered as separate rather than integrated tasks.

The construction of the dam was completed in 1995, and the 500 displaced families were compensated, moved out of the area completely, and resettled in new areas, some as far as 500 km away. In 1994, NCC commissioned two consulting groups to carry out research to evaluate the implementation of the environmental-action and -monitoring program drawn up in 1989, as well as any other NCC actions during the dam construction, either around the dam or elsewhere in the project area. In addition, the consultants were to make further recommendations to improve the project's sustainability and develop management options for the reservoir and the route followed by the tunnels and pipelines over the 60-km distance between the dam site and the distribution point in Nairobi. Unlike previous studies, however, in this study the World Bank and NCC made a deliberate effort to have integrated research on the two major themes, the environmental and socioeconomic impacts of the project.

Management of the evaluation research

Research planning

Two consulting teams were commissioned to carry out the research, one to evaluate the socioeconomic impacts of the project and the other to evaluate the environmental impacts. The consultants had earlier participated in preparing the 1989 action plans. NCC initially invited them to submit their proposals to carry out the evaluations. The consultants were then invited to a meeting with representatives of NCC and the World Bank to discuss their proposals in detail. With the necessary amendments, the proposals formed the terms of reference for the work, as two separate projects, to be carried out with appropriate collaboration between NCC and the researchers. Although it was unclear to both the researchers and NCC how best to collaborate in this project, it was clear from the terms of reference that the results were to lead to integrated-policy development. The main terms of reference for the researchers were the following:

+ To evaluate actions taken to contain disturbances in the dam catchment and their ecological impacts;

+ To evaluate actions taken to contain disturbances and soil erosion along the pipeline route;

+ To assess and evaluate the sludge-waste disposal system at water-treatment works and suggest remedial measures to counteract the increased sludge–water ratio.

+ To determine whether displaced households were better or worse off than they had been before the project, in terms of access to social amenities (education, health, transport, etc.) and economic status;

+ To evaluate the effectiveness of the resettlements and the rehabilitation process for displaced households;

+ To recommend future management options for the reservoir and pipeline route, taking care to recognize that only environmentally friendly activities could be carried out by either NCC or the local community; and

+ To work in collaboration with appointed NCC officials, to submit the draft report within 3 months for comments from NCC and the World Bank, and to submit a final report within 1 month of the receipt of comments from NCC and the World Bank.

Vertical and horizontal integration

To facilitate the consultants' work, NCC assigned a sociologist in charge of the resettlement program to the socioeconomic team and a water chemist to the environment team. The two officials held a joint meeting with the consultants to agree on a preliminary work plan.

The consulting teams each had members from the University of Nairobi. The lead consultant for the socioeconomic study, a land economist specializing in real-estate valuations and compensations, formed a team with a sociologist, an economist, a public administrator, and 10 graduate assistants. The lead consultant for the environmental study, an ecologist, formed a team with a zoologist, an agriculturalist, a water-quality chemist, a public-health technician, and a laboratory technologist. The two lead consultants, who became the team leaders, then independently organized separate work programs. Each team then separately met with its respective NCC official — a departure from the integrated-policy-research process initially set out by NCC and the World Bank — and agreed on detailed work plans.

The socioeconomic team used questionnaires, observations, and sampling to survey the displaced households, scattered in five districts of the country (it had been impossible to secure enough land for a single resettlement site). The NCC sociologist, in collaboration with the provincial administration, helped the team trace 259 household heads (of the 500 displaced) and their dependants. The team conducted interviews, as well as made observations, on how the families were coping with life in their new homes. In addition, the team interviewed a sample of officials from NCC, the central government, the contractor, and people living in the vicinity of the project.

The questionnaires had been drawn up with sections dealing with different issues, according to the prime areas of concern of each specialist. After discussing the sections together, the research team and NCC arrived at a consensus about the sequence and wording of the questions. The team split up into groups consisting of one researcher and three or more research assistants to administer the questionnaires. Results from separate groups were entered and analyzed together using Statistical Package for the Social Sciences software.

The economist basically did the quantitative analysis and presented the results. Presenting the qualitative aspects was mainly the work of the sociologist and the public administrator, who took notes during the field surveys on their impressions of the respondents and, together the with lead consultant, carried out most of the background literature review. Although the report was written in parts by four consultants, using the appropriate data, it was edited by the lead consultant and the sociologist. The typed draft was then circulated to the four researchers and NCC for their comments.

The lead consultant, after receiving the comments, called a meeting of the researchers and the NCC official and circulated the comments for discussion. The lead consultant then prepared a final draft of the report and submitted it to NCC to be incorporated with the report from the environment team.

Contrary to expectations, not much collaboration took place between the two research teams during data collection, analysis, or report writing. Their methods of investigation, as well as their perceptions of the project, were different in the main. One team was more concerned about people, whereas the other was more concerned about the environment.

The environment team used scientific laboratory techniques and observations in their research. Unlike the socioeconomic team, which traveled widely, this team concentrated its work in the project area, where it observed the eco-anthropological changes that had occurred in land-use patterns in the watershed and adjacent catchment areas and the impacts of construction activities on the riparian ecosystems, from the dam site to the treatment plant, and beyond, to the treated-water reservoir in Nairobi. This team also collected samples of water from the streams for quality testing (dissolved oxygen, temperature profiles, etc.) and to test the fishery potential of the dam. Like the socioeconomic team members, each member of the environment team had a separate assignment, to be carried out independently within a given time frame. Each researcher drafted a separate piece, which was submitted to the lead consultant for discussion and subsequent editing.

When reports from both teams were ready, NCC called a meeting to discuss the subreports and the best way to incorporate them into one document. There was also the issue of who was to merge the two reports. It was agreed that the draft reports, "Socio-economic Appraisal," and "Physical and Environment Evaluation," be submitted in their present forms to NCC and the World Bank for their comments. After receiving the comments from the two institutions, the lead consultants edited the final reports and presented them in two separate volumes.

Again contrary to expectations, the two reports were not jointly written or cross referenced. Instead, they have to be read as independent reports. This makes integration in policy formulation and implementation difficult. The fault here lies with the two research teams, which considered their assignments as being different and independent, as well as with NCC, which had neither the idea nor the experience of coordinating integrated research.

Assessment of the research

The research was policy-demand driven and made useful recommendations to the NCC and the Kenyan government on how best to handle compensation and resettlement to avoid human misery and how best to manage the dam and the pipeline in an environmentally friendly way. As discussions with NCC officials showed that they were enthusiastic about putting some of the recommendations into practice and thereby creating a good research–policy nexus, it is unclear why the proposed action plans were never fully implemented.

In terms of integrated-policy research as theoretically conceived, this research project fell short of some of its requirements. The role of the beneficiaries, for instance, was neglected during the research design, as well as during the preparation of the report. The beneficiaries should have included those displaced, those around the dam site and pipeline, and Nairobi residents who receive the water. The management of the reservoir and the pipeline requires the participation of the community to prevent water pollution, avoid damage to the pipes, etc. The beneficiaries only participated as interviewees and had no ownership of the research in any sense. For this reason, unless NCC provides incentives, it may be difficult to use the research results in environmental-protection initiatives. In other words, appropriate vertical integration was not incorporated into this study, even under the conventional participatory research. At the time of the survey, this resulted in residents along the pipeline puncturing it at various points to draw water, which, if uncorrected, may lead to a lot of wasted and unaccounted for water.

With respect to horizonal integration, the conduct of the research could be described as multidisciplinary, rather than interdisciplinary. Although the research was multisectoral in terms of involving several disciplines, the process lacked collaboration, cooperation, and communication between the two teams. Some form of cooperation and communication occurred, but only within each team. The effect of this was duplication in collection of some socioeconomic data, such as on changes in the land-use patterns of people whose land was only partly acquired, like those along the pipeline wayleave. This information could have been collected by only one group if the two groups had compared notes initially about their research methods. In some instances, the researchers contradicted each other if observations had been made at different points. For instance, whereas the socioeconomic group observed that the contractor had not rehabilitated most of the land along the pipeline, the environment group gave the impression that the contractor had rehabilitated the land by planting Nippier grass. In fact, the landowners had done the rehabilitation, and they have since asked NCC for additional compensation for work carried out on behalf of the contractor. Such contradictory findings, of course, led to contradictory recommendations for policy.

The final report reflects the fact that the synthesis of the data was more an accumulation of various findings than an integration of the research components. This is especially noticeable in the environment volume, with each part reading independently of the others. The socioeconomic team at least had one survey instrument and methodology for data analysis.

The two volumes of the report were edited for uniformity of writing style. Although each volume maintained the scientific rigour of its respective discipline, the whole report failed to exploit the commonalities. The fault lies in the professional prejudices of the researchers, as well as in NCC's lack of expertise in coordinating integrated research.

Finally, the research failed to provide any training or workshop to involve stakeholders. The reports were presented in writing only to the researchers, NCC, the World Bank, and the government. The public and other institutions were unaware of the findings and the recommendations. Because the research was commissioned, NCC and the World Bank now claim to own the information and researchers may not disseminate it without their authority. However, permission may be possible if the researchers are willing to disclose only the positive aspects of the findings.

Despite its shortcomings, the evaluation exercise did reveal a number of positive aspects. Indeed, as Syagga and Olima (1996) observed, the Third Nairobi Water Supply Project had both positive and negative impacts. On the positive side, the project created direct employment at the construction site and in small-scale businesses that serve the construction workers. Labour for future dam maintenance will also be required. The project has also created a number of facilities for the benefit of the surrounding community, notably a primary school, a health clinic, an improved road network, and a power-supply station. Above all, the project will supply enough water to sustain Nairobi's growing needs to 2005.

However, the project has caused a considerable amount of suffering for displaced households. Extended families were split up, compensation money was inequitably distributed within households, the total compensation was insufficient, and families suffered a loss of earning capacity. The dam altered the microclimate in the vicinity, and more mosquitoes are breeding and causing sickness in the surrounding areas. Farmers are also having to deal with soil erosion arising from the unprotected high embankments, and the remaining parcels of land are economically unproductive. These should have been considered during the feasibility study, which points again to the need for integrated-policy research.

Research–policy links

Vertical integration in the project was weak. The participatory role of the community members was limited to that of respondents rather than that of owners of the project. To date, they have no idea what the research findings were or what role they would be expected to play in the future.

At NCC, only the Water and Sewerage Department officials were involved. No other NCC departments or any of its policymakers participated, so they may be unaware of the relevance of the research and may not support the implementation of its recommendations, particularly any that have financial implications. A dissemination seminar or workshop would therefore have been most appropriate for the rest of NCC's officials, whose future support may be very crucial. Some of the recommendations touch on national policy, such as land-acquisition processes and compensation and resettlement of displaced populations. Other recommendations, like using the dam for sports and fishing, will need collaboration from the Ministry of Environment and Natural Resources or the Ministry of Tourism for implementation. These ministries and many others need to be familiar with the project and have an opportunity to give their opinions and support in advance of implementation. NCC therefore requires a lot of political goodwill, as well as professional support, from various organs of the central government to implement the recommendations. Equally important is the need to sensitize other local authorities and institutions in the country that may undertake similar projects so that they can benefit from the lessons already learned. The dissemination methods adopted so far have, however, precluded any meaningful participation in this research–policy link on a wider scale.

Thus, some of the major shortcomings of the research–policy links in this project can be summarized as lack of community participation, ineffective reporting and dissemination strategies, and inappropriate institutional structures for implementation. However, the impact of the study is gradually gaining ground. The two lead consultants were subsequently invited to participate in a proposed major hydroelectric generating project, principally to deal with issues of land compensation and EIA. However, even in this research project, the issues of vertical integration, particularly with respect to community involvement, remained unresolved.

Conclusion

The evaluation research for the Third Nairobi Water Supply Project is an example of the integrated research typical of many African countries. The research had socioeconomic and environmental components. It was policy oriented and conducted by researchers from a number of disciplines in the social and natural sciences. The results had important implications for policy formulation both at the national level and at the specific-project level. The role of institutions in promoting an interdisciplinary understanding of issues was evident, particularly when NCC decided to award the consultancy as a joint project (although with two separate contracts). The leadership appeared effective, leading to timely completion of the research. The incentive system, in the form of consultancy fees, appeared appropriate.

However, in East Africa, the concept of integrated-policy research is yet to be appreciated. This example of a research project exhibits a number of weaknesses in the ways integrated-policy research is conducted that may be typical for the region. Full participation of the beneficiaries was lacking in the research process. Other organs of NCC and of the central government that were crucial to policy formulation were also uninvolved. This suggests a lack of appropriate mechanisms and institutional structures to facilitate research–policy links. Poor horizontal integration between the research teams resulted in accumulation of various findings, rather than the integration of research components. The methods of reporting and the dissemination strategies appeared ineffective for wider research–policy links.

Although data collection by the socioeconomic team created no serious problems, some issues arose, such as how better or worse off a particular household was and what would be adequate social compensation. No consensus was achieved on the assessment of nonuse values of the land because these are not easily quantifiable. The environmental team had even greater problems with indicators of environmental quality. For instance, Kenya has no appropriate pollution standards, thus making it difficult to implement policies on conservation and pollution control and monitor their effects (Syagga 1994). Even in the developed countries of the North, indicators of sustainable development are the subject of continuing debate in economic planning and environmental management. Various models are proposed and continually revised, using simulations (Kuik and Verbruggen 1991). Adequate experience is also lacking in mobilizing popular participation in the formulation and implementation of policies at the national level. In this respect, integrated-policy research has a role to play, as research is about generating new knowledge and solving problems and can help policymakers to either select interventions for specific contexts or improve interventions.

Recommendations

This case study provides some key lessons for future research efforts aimed at integrating policies for sustainable development in Africa:

+ Research problems should be identified in close consultation with the users of the research results. If research problems are improperly identified, then research on a problem is likely to produce findings irrelevant to the user, creating bottlenecks in research–policy links. A more recent feasibility study for a hydroelectric generating project in Kenya, being carried out by two leading engineering firms, is facing similar problems because of limited consultation with the people whose property will be acquired for the project.

+ Monitoring and evaluation should be part of the research. Research is conducted to find solutions to specific problems or to better understand a given situation. Without a process of monitoring and evaluation, it is difficult to determine how successful the solutions are. What is more pertinent, however, is the timing of the appraisal. In this case study, for instance, elaborate action plans were to be implemented in 1989, but the evaluation, which was not carried out until 1994, showed that implementation was still incomplete. Perhaps it was too late or too costly to correct some of the negative impacts.

+ Interdisciplinary research projects should be followed by policy-formulation workshops. Government planners should be encouraged to participate in and use such workshops to develop their own ideas on how to cope with policy options. Government departments should also be encouraged to work with scientists on live problems to strengthen the link between the researchers and the policy planners and implementors. However, the amount of funding allocated for policy research is usually such a small proportion of any total project funding that it rarely allows for dissemination workshops.

+ Research and training need a more integrated approach. Research projects need to provide or to be integrated with planned training components that build up the necessary skills of researchers, technicians, and support staff in training institutions. A few institutions, such as the School of Environmental Studies at Moi University in Kenya, are now adopting integrated research and training in their programs.

+ National governments need to formulate environmental policies for both urban and rural settlements. EIA requirements for proposed development programs imply that environmental indicators and standards have already been developed. This is far from reality in most African countries. Accordingly, national governments and the international community should support and encourage research in environmental management to provide baseline data for formulating polices. A few countries, notably Kenya, Tanzania, and Zimbabwe, have national environmental action plans, but implementation is still hampered by a lack of scientific parameters. For instance, countries like Kenya and Tanzania, which have traditionally harboured refugees, have, to date, been unable to appreciate and accordingly to plan for environmentally sustainable refugee settlements.

✦ Institutional capacity of various national governments and related agencies needs to be strengthened so that they can cope with the needs of integrated-policy research. Just as sustainable development requires that environmental costs be integrated into economic systems, it also requires that citizens be integrated into the political process. The pursuit of sustainable development requires a political system that recognizes effective citizen participation in decision-making. The national environmental action plans, which are in various stages of development in many African countries, should recognize this requirement.

Chapter 8

Sustainable Irrigation in the Arid Regions of India

V.S. Vyas

✧ ✦ ✧

Introduction

This case study examines a development strategy for agricultural growth, ecological sustainability, and socioeconomic equity in a sparsely populated arid
region served by a major irrigation project. The study focuses on the command
area of the Indira Gandhi Canal Project (better known by its Hindi name *Indira
Gandhi Nahar Pariyojana*, IGNP) in northwestern Rajasthan, India.

Regional context and development challenges

Socioeconomic profile

Rajasthan is one of the developing states of India. Its average per capita income
in 1991/92 was 1 717 INR (in 1998, 39.55 Indian rupees [INR] = 1 US dollar), far below the national average of 2 250 INR. In terms of per capita income,
it was ranked tenth (in descending order) among the 16 major states of India
(CMIE 1992). In terms of infrastructure development, as well as of overall
human-resource development, the state occupies a place close to the bottom (see
Tables 1 and 2).

Rajasthan's economy is based on agriculture and animal husbandry.
Agriculture accounted for 50% of net state domestic product in 1988/89.
Nearly 69% of those in the work force depend on agriculture and work as cultivators or agricultural labourers. Crop yields are generally low, and the further
growth of agriculture is constrained by a lack of water and access to it, as nearly
60% of the region is arid. Even outside the arid region, agriculture depends on

Table 1. Index of relative development of infrastructure, 1966–93.

State	1966/67	1976/77	1980/81	1985/86	1986/87	1987/88	1988/89	1989/90	1990/91	1991/92	1992/93
Andhra Pradesh	93	97	98	105	104	103	100	101	98	103	103
Arunachal Pradesh	—	—	—	—	—	43	30	31	32	43	44
Assam	73	89	93	87	96	93	91	95	95	94	93
Bihar	98	104	97	98	99	98	96	96	97	96	96
Goa	—	—	—	—	—	150	128	128	109	144	171
Gujarat	111	122	125	132	132	130	128	125	124	122	125
Haryana	129	151	154	150	149	148	148	149	156	153	152
Himachal Pradesh	—	72	79	86	85	84	86	85	86	87	84
Jammu and Kashmir	83	77	73	70	74	73	72	71	70	72	69
Karnataka	90	105	101	100	100	98	95	99	93	99	97
Kerala	135	167	137	140	140	140	137	140	138	139	140
Madhya Pradesh	53	61	62	71	71	72	72	72	72	72	75
Maharashtra	117	111	118	119	118	114	113	112	111	110	111
Manipur	—	63	73	78	77	80	76	78	78	79	81
Meghalaya	—	63	60	84	71	70	70	71	69	66	65
Mizoram	—	—	—	—	—	63	57	57	63	64	63
Nagaland	—	81	77	77	71	72	69	67	73	71	71
Orissa	69	79	82	81	83	82	82	82	86	88	89
Punjab	201	216	215	218	216	214	210	214	211	210	205
Rajasthan	59	81	77	79	79	78	76	82	85	81	80
Tamil Nadu	171	152	153	142	142	142	137	139	139	138	138
Tripura	—	48	55	80	62	61	61	66	65	63	63
Uttar Pradesh	107	112	107	108	108	107	106	108	111	109	109
West Bengal	152	133	132	123	123	121	116	116	115	115	113
Ratio of highest to lowest (major states)	3.8	3.5	3.5	3.1	3.0	2.9	3.0	2.9	2.9	2.9	2.7
SD from all-India average	40	39	37	36	35	34	34	34	34	33	32

Source: CMIE (1992, 1994).
Note: SD, standard deviation.

Table 2. Ranking of Indian states according to human development.

State	Literacy, 1991	Female literacy, 1991	Infant mortality, 1988–90	Life expectancy
Andhra Pradesh	11	11	7	7
Assam	9	8	11	14
Bihar	15	14	10	11
Gujarat	4	5	9	8
Hariyana	8	9	8	3
Karnataka	7	7	6	4
Kerala	1	1	1	1
Madhya Pradesh	12	12	14	13
Maharashtra	3	3	3	5
Orrisa	10	10	15	12
Punjab	6	4	2	2
Rajasthan	14	15	12	10
Tamil Nadu	2	2	4	6
Uttar Pradesh	13	13	13	15
West Bengal	5	6	4	9

Source: GMP (1995).

rainfall, which in most parts of Rajasthan is minimal and irregular. Only 24% of the cropped area is under irrigation, and only half of the irrigation potential is being used (CMIE 1994).

The state has a lower population density, at 128 persons/km^2 (in 1991), than the country does, with 267 persons/km^2. The average size of agricultural holdings, 4.34 ha, is much larger than the national average of 1.68 ha. A much larger proportion of the cultivators in Rajasthan are owner–operators. Thirty-four percent of the population lives below the poverty line (Hashim 1995). Although this is below the national average (39%), it poses a significant development challenge to Rajasthan, given the low productivity of land and limited availability of water.

Agroclimatic profile

The State of Rajasthan is divided into two geographical regions: the Aravali Range, traversing the state from northwest to southeast, acts as a sharp divide; the region west of Aravali is the extension of the Thar Desert. IGNP is in this latter region. This mostly arid and partly semiarid region is sparsely populated, but the density of livestock is very high. The crop yields are low and show sharp year-to-year variation. From the southeast to the northwest in this region, the agroclimatic conditions become progressively harsher. In the part covered by stage II of IGNP, the aridity is highly pronounced. The landscape is dotted with numerous sand dunes. Many of these shift from place to place with the brisk winds of the desert, leading to desertification of the adjoining areas. This region is also characterized by extremes of climate, erratic rainfall, and high evapotranspiration and is subject to recurring droughts.

Agricultural-development policy

Rajasthan has undertaken several measures for agricultural development, with expansion of irrigation being the most important of these. The *Indus Water Treaty* settled the long-standing dispute between India and Pakistan about sharing the waters of the Punjab rivers. The waters of three eastern rivers — Ravi, Bias, and Sutlej — were allotted to India. From India's (annual) share of water, Rajasthan got 8.60 million acre–feet (MAF; 1 acre–foot = 1.33 ha–m). The Government of Rajasthan decided to use 7.59 MAF in the construction of the Indira Gandhi Canal. The remaining water in Rajasthan's share is used for the Bhakra and other irrigation systems. Of the total water supplies in IGNP, 6.72 MAF is earmarked for irrigation; 0.87 MAF, for drinking water and industrial use in the command and adjoining areas.

In 1956, the end of the First Five-Year Plan, only 12.7% of the gross cropped area in the state was irrigated. By 1990, the end of the Seventh Five-Year Plan, the proportion had increased to 24.9% (GOR 1992), mainly as a result of IGNP. At the same time, Rajasthan adopted a high-yielding-varieties program as part of its agricultural strategy. Other developmental measures included investment in agricultural infrastructure and marketing, increased credit, improved research capability, and increased input supplies. Such measures led to a growth in agricultural output of 4.68% per year during 1980–91.

The government's development strategy for the area covered by stage II of IGNP had two aims: irrigation development and human settlement. Stage I of IGNP primarily provided the rationale for this strategy. Physical and demographic conditions in the stage I districts, Ganganagar and Bikaner, were more or less similar to those prevailing in the stage II area. With the introduction of irrigation and planned human settlement, the economy of the area under stage I of IGNP had undergone a remarkable change (WAPCOS 1992b). It was assumed, therefore, that these two components of development strategy would yield similar results in the stage II area.

However, people in some quarters raised serious doubts about the suitability of the arid lands, termed "fragile lands," for intensive agriculture and challenged the use of irrigation on these lands, on the grounds that the light sandy soils of the desert may be unsuitable for irrigated farming or for any type of intensive cultivation. Also, large parts of the region have hardpan at shallow depths. Irrigated farming under these conditions increases the risk of soil erosion, desertification, water logging, and salinization (Ramanathan and Rathore 1994). These objections were overruled by the state government. It pointed out that improvement in the vegetation cover and the extension of afforestation (which would be possible with the supply of water assured through irrigation) would stabilize sand dunes and halt the process of desertification. Similarly, the occurrences of hardpan at shallow depths would call for a more judicious use of water, which was seen as a blessing in disguise.

The project

Because of the extent of poverty in the region, the heavy reliance on agriculture, and the limited availability of water, the state government saw in irrigation a significant potential for increasing agricultural production and, in turn, alleviating poverty in the region.

At 649 km long, IGNP is one of the world's biggest canal projects. It takes off from Sutlej River at Harike Barrage and runs 204 km as a feeder canal, before it enters Rajasthan. Within Rajasthan, the canal is 445 km long. IGNP covers seven districts of Rajasthan: Barmer, Bikaner, Churu, Hanumangarh, Jaisalmer, Jodhpur, and Sriganganagar. The salient features of the project are given in Table 3.

Stage I of the canal, which involved the construction of 204 km of feeder canal and 189 km of the main canal, was started in 1957 and completed in 1973. Stage II (256 km of the main canal) was completed in 1986. By 1997, the construction of distributaries and the development of the command area had also progressed (Table 4). The analysis in this case study concerns the aspects of sustainable and equitable development in the stage II area of IGNP.

Broadly speaking, the socioeconomic and biophysical objectives of the irrigation project were to

+ Halt the process of desertification;

+ Develop agriculture in the command area;

+ Create human settlements in the sparsely populated area; and

+ Provide drinking and industrial-use water to the inhabitants of the project area and adjoining regions.

Table 3. Salient features of the Indira Gandhi Canal Project.

	Stage I	Stage II
Feeder canal (km)	204	—
Main canal (km)	189	256
Distribution system (km)	2 950	5 115
Gross command area ($\times 10^5$ ha)	9.5	15.6
Cultivatable command area ($\times 10^5$ ha)	5.25	10.12
Flow area command ($\times 10^5$ ha)	4.79	7.00
Lift area command ($\times 10^5$ ha)	0.46	3.12
Head discharge (cusec)[a]	18 500	9 955
Water allowance (cusec)[a]	5.23	3.00
(flow lift/1000 acres)[b]	4.30	3.00
Intensity of irrigation (%)	110 (37K, 63R)	80 (35K, 45R)
Water allocation (MAF)	3.59	4.00
Drinking water (MAF)	0.22	0.65
Irrigation water (MAF)	3.37	3.35

Source: WAPCOS (1992a).
Note: K, *kharif* (summer season); R, *rabi* (winter season); MAF, million acre–feet (1 acre–foot = 1.33 ha–m).
[a] 1 cusec = 1 ft^3/s (= 0.028 m^3/s).
[b] 1 acre = 0.405 ha.

Table 4. Progress of canal construction and command-area development, to March 1996.

	Stage I		Stage II	
	Achieved	Remaining	Achieved	Remaining
Canal works				
Main canal, including feeder (km)	393	Nil	256	Completed
Distribution system (km)	3 188	212	2 447	3 333
Cultivatable command area ($\times 10^3$ ha)	531	22	509	807
Irrigation potential ($\times 10^3$ ha)	531	22	—	—
Irrigation achieved, 1992/93 ($\times 10^3$ ha)	620	—	130	729
Investment ($\times 10^6$ INR)	3 210	980	11 020	24 200
Command-area development				
Construction of watercourses ($\times 10^3$ ha)	442		285	1 031
Construction of roads (km)	1 066		899	1 188
Sanitary diggies (n)	219		153	380
Pasture development, etc. ($\times 10^3$ ha)	94.7			
Canal-side plantation (Rkm)	8 015			
Road-side plantation (Rkm/10^3 ha)	3 385/2.6			
Afforestation ($\times 10^3$ ha)	—		58	84
New settlers (n)	94 908		44 659	—
Area allotted to new settlers ($\times 10^3$ ha)	719		284	746
Investment ($\times 10^6$ INR)	2 862		5 579	10 021

Source: IGNP (1996).
Note: INR, Indian rupees (in 1998, 39.55 INR = 1 US dollar); Rkm, running kilometre.

The following subsections focus on the first three of these four objectives, as well as on the market and institutional failures affecting their attainment.

Halting the process of desertification

According to some analysts, the introduction of agriculture into arid areas is ecologically unsustainable because, by definition, arid lands are fragile. Irrigation invariably leads to intensive agriculture, which cannot be sustained on these lands. These lands are best suited for extensive agriculture, mainly pasture-based animal husbandry (Ramanathan and Rathore 1994). According to this way of formulating the issue, fragile lands are loosely defined and are equated with marginal lands. Two characteristics distinguish fragile land: environmental sensitivity (propensity to deteriorate) and resilience (ability to retain productive qualities) in the context of common land use (Turner and Benjamin 1994). Historically, arid lands have sustained intensive agriculture, given an assured and renewable supply of water. Proper land management and land-related investment can overcome ecological constraints. These lands are unsuitable for crop production only if farmers have to rely exclusively on (scanty and irregular) rainfall or an exhaustible source of water, such as groundwater.

On the other hand, extensive farming, particularly animal husbandry, has placed excessive pressure on land resources, aggravating the arid conditions. Excessive grazing, in addition to damaging the vegetation cover, loosens the sandy soils; the winds then shift the sand to adjoining areas, accelerating desertification. This is not an argument in favour of monocrop or Green Revolution agriculture but suggests that, with dependable irrigation, an integrated-farming system — comprising crops and livestock in tune with the local resources and the farmers' needs — could be sustainable on arid lands (Jodha 1990). In such a case, irrigation would promote environmental sustainability on arid land. The experience with irrigation in arid areas of India and abroad supports this conclusion.

Two examples of such experience can be cited from the arid regions of Rajasthan. The most famous and most pertinent is that of Gang Canal. In 1927, the ruler of the State of Bikaner thought of bringing Sutlej waters to the parched arid land in the north of the state and constructed the Gang Canal in the area, which is now known as Sriganganagar (Sain 1978). The state followed a judicious settlement policy. Today, Sriganganagar is agriculturally the most progressive district of Rajasthan. Its yield rates in important crops, like wheat and cotton, are comparable to those of the Punjab. Agricultural prosperity diversified the economy of the district, which now ranks fourth among 31 districts of the state in terms of industrial development.

This story is being repeated in the region covered by stage I of IGNP. The rate of growth in agricultural production in this region is now close to that of Sriganganagar. The economy is diversifying with the growth of agroprocessing industries. A study by the National Council for Agricultural Economic Research showed that the population is reaping the social and economic benefits of irrigation (Roy 1983). More important, canal irrigation has resulted in large-scale afforestation, especially canal-side afforestation, and, to a limited extent, roadside afforestation. This has affected the microclimate, and erosion has slowed as a result of the increased humidity, moisture, and vegetation cover.

"Waterlogging" and salinization pose the real threat to the environment in the IGNP area. A number of studies revealed that the lands to be irrigated by IGNP in stage II have low groundwater levels (WAPCOS 1992b). Furthermore, hardpan layers at shallow depth allow no water to permeate downward. Together with the capillary action, this leads to water oozing. Salts in the desert soils further complicate the situation, leading to salinization. However, the experts differ about the extent and severity of these problems. An alarmist view suggests that water is rising at the rate of 0.8–1 m/year in large parts of the stage II area and that nearly 34% of the area is likely to become waterlogged (WAPCOS 1992b). On the other hand, experts on the *Indira Gandhi Nahar* Board (IGNB) heavily discount these estimates and suggest that no more than 2–4% of the stage II area is truly vulnerable. A more reasonable view is that the proportion of area affected is neither as high nor as low as the official figures indicate, although the rapid spread of waterlogging is distinctly possible unless effective steps are taken.

In addition, water-management practices can either aggravate or relieve the situation. At the moment, these practices are accelerating the waterlogging;

seepage of water from irrigation channels and watercourses and excessive, non-judicious use of water are aggravating the situation. The way to tackle this problem, apart from correcting the faults in the water-conveyance systems, is to reduce the water allowance and change the cropping pattern so that, even with the reduced supply of water, farmers can grow remunerative crops.

In any case, action to reduce the water supply per hectare of land will have to be taken soon because more areas are developing and the fixed quantity of water will have to be made available to larger areas. Initially, the project provided for a water-use intensity of 110% in the stage I and II areas, but this was later reduced to 80% in the stage II area. So far, it has been impossible to keep to these standards because extension of the irrigated area in stage II has been slow. As a result, the irrigable area in stage I has been getting more than its share of water, with the irrigation intensity rising to 130%. With the water-conveyance system in place in the stage II area, water will have to be spread more evenly and it should be possible to reduce the supply of water to those areas susceptible to waterlogging. Experience during the drought years in the IGNP area shows that if the supply of water is reduced, the water table also goes down and, to that extent, so does the risk of waterlogging (IGNB 1995). A deliberate plan to reduce the water allowance may lead to better water management and less risk of waterlogging and salinity.

To achieve the water-management objectives, three hurdles have to be overcome: problems related to the political economy of the area, problems related to management of demand, and technical problems. One of the seriously waterlogged areas was adjacent to the Ghagghar Depressions in the stage I area. Initially, the water was diverted and released in these depressions, situated along the banks of Ghagghar River, to save the lands on the banks from the ravages of flood. These lands are owned by the rich and politically well-connected people, whereas the villages adjacent to the Ghagghar Depressions are populated by resource-poor farmers. A permanent liability has been created for the poor farmers by saving the lands for more affluent people. The situation is aggravated because the surplus water in the Ghagghar Depressions is kept in pools to enable the development of fisheries. These fisheries are also owned by the richer farmers of the area (Srivastava and Rathore 1992). Accumulation of water in these depressions contributes to waterlogging of low-lying lands, which were at one time productive.

A more difficult problem is that a large number of farmers in the stage I area have grown accustomed to a water regime that enables them to obtain an irrigation intensity of 130% or better. Would they be content to live with an irrigation intensity of, say, 100–80%? Answers to such questions hinge on demand-management techniques, such as

+ Organization of different sections of people in these areas;

+ The extent of the possible loss by the farmers in the stage I area when the existing water regime changes;

+ Possibilities of compensating for these losses (say, with higher-yielding crops that use less water) and of having a general system of incentives and deterrents to induce farmers to move to water-saving cropping patterns; and

+ The capability of the state and the bureaucracy to enforce the equitable distribution of water.

Technical challenges include the following:

+ Finding ways to drain excess water;

+ Adopting agronomic practices and cropping patterns that use less water;

+ Fixing the water allowance for different areas, with a view to maintaining soil characteristics and groundwater availability; and

+ Building technically sound structures and systems of water conveyance.

The success of these interventions depends largely on the involvement and participation of the beneficiaries, which is an aspect repeatedly overlooked by project-planning authorities.

Developing agriculture

The second broad objective of IGNP was to ensure high levels of agricultural production in the command area. Irrigation of 950 000 ha in stage II would yield agricultural products worth an estimated 20 billion INR. The project would result in enough annual fodder for 5.2 million cows or equivalent animal units and facilitate afforestation of 362 000 ha, producing 7.16 million INR of forest products annually (IGNB 1995).

The main production gains should be in crop yields. However, from the available information, the actual yields are reaching only part of their potential. For example, actual yields of different crops per hectare, compared with the potential yields of the same crops, vary from 31% for gram to 60% for cotton (see Table 5). Although undue importance should not be attached to the figures of achievable and potential yields, it is quite clear that further improvements in the yields of various crops are possible.

For irrigation to make an impact on agricultural productivity, several conditions have to be satisfied:

+ A dependable supply of water at source;

+ Creation of irrigation potential at the field level;

+ Efficient conveyance of water;

+ Appropriate cropping patterns; and

+ Complementary inputs of improved seeds, fertilizers, etc.

Table 5. Reported yield and achievable yield in the command area.

| | Yield (quintals/ha) | | | | | |
| | Kharif | | | Rabi | | |
	Cotton	Guwar	Groundnut	Wheat	Gram	Mustard
1984/85	13.38	8.31	18.14	13.32	6.15	10.45
1985/86	11.38	9.53	13.61	20.00	8.00	12.00
1986/87	12.74	7.46	15.93	20.39	5.64	8.35
1987/88	5.45	4.98	7.72	12.29	3.62	6.42
1988/89	11.17	8.73	14.20	30.33	7.75	12.34
Average	10.82	7.80	13.92	19.26	6.23	9.91
Achievable	18.0	17.5	25.0	35.0	20.0	18.0
Actual as % of achievable	60	45	56	55	31	55

Source: WAPCOS (1989); IGNP (1995a).
Note: *Kharif*, summer season; *rabi*, winter season.

The first hurdle is the fluctuation in the supply of water at the canal head (based on fluctuations in the supply of water from the rivers). Added to this are the multiple claims of various riparian states on the given supply of water, normal difficulties faced by the tail-end states, and the seasonal variability in the availability of water at the canal head. Because Rajasthan has no reservoir, mechanisms to even out the flow of water to the command area need to be developed. For example, adopting a rotational scheme among the distributaries and developing a *varabandi* system (a scheme whereby water is allocated on a weekly basis) at the outlet would allow for more equitable (but not necessarily more efficient) distribution of the water.

The second challenge is to create irrigation potential (which is basically a function of the construction of the main canal, distributaries, and field channels). From Table 4, it is clear that the creation of potential has been slower in the stage II area than in the stage I area. This is partly because of the difficult terrain and, to some extent, less-experienced farmers in the stage II area. The reasons for the slow progress in bringing the cultivatable area under actual irrigation have to be examined before the full production potential can be realized.

A third factor inhibiting the maximization of agricultural production is inefficient water conveyance. This is mainly due to problems of maintenance, accumulation of sand in canal distributaries and outlets, and unchecked growth of weeds. Water-conveyance efficiency in stage II is reckoned to be 72.6% (Ramanathan and Rathore 1994). With the defects in the conveyance system, the availability of water on the farm is less than it is presumed to be at the macro level.

The cropping pattern and agronomical practices in vogue in the IGNP area create excessive use of water (Table 6). Crops requiring a lot of water, such as wheat, are prominent. However, the number of irrigation canals is much higher than agronomically justified; the actual water requirement is lower than the amount provided to the farm, even after all conveyance losses are taken into account. In fact, in two distributaries where the Command Area Development

Table 6. Areas opened by canals, covered by watercourses, and actually irrigated in stage I and in the corresponding periods in stage II.

		Area ($\times 10^3$ ha)		
		Opened by canals	Coverage by watercourses	Actually irrigated
Stage I				
1	1974/75	286	0	258
2	1975/76	347	1	289
3	1976/77	403	35	279
4	1977/78	445	64	294
5	1978/79	461	114	322
6	1979/80	471	139	348
7	1980/81	476	153	360
8	1981/82	486	207	402
9	1982/83	505	238	426
10	1983/84	515	288	438
11	1984/85	514	328	416
12	1985/86	515	361	463
13	1986/87	517	370	524
14	1987/88	519	395	375
15	1988/89	522	408	553
16	1989/90	524	410	524
17	1990/91	525	411	578
18	1991/92	525	411	577
19	1992/93	525	416	615
20	1993/94	525	418	552
21	1994/95	529	432	617
22	1995/96	531	442	660
Stage II				
1	1986/87	95	0	5
2	1987/88	124	1	7
3	1988/89	145	30	12
4	1989/90	187	90	17
5	1990/91	232	142	40
6	1991/92	278	182	60
7	1992/93	322	214	71
8	1993/94	386	230	84
9	1994/95	447	248	102
10	1995/96	509	285	130

Source: IGNP (1996).

Authority ensured low but regular water supply, it was found that better yields could be obtained by a smaller supply of water (Ramanathan and Rathore 1994).

In sum, several factors conspire to inhibit the full exploitation of the yield potential of various crops. To remedy all these deficiencies, action is required from several actors, but especially from the settlers in the area.

Creating human settlements

As mentioned earlier, the area coming under the command of IGNP in stage II was sparsely populated; in fact, it had a population density of 11 persons/km². Very little agriculture was undertaken in the area; the main occupation of the inhabitants was animal husbandry. With the introduction of irrigation and the opening up of land, a large number of settlers had to be attracted to the area to make use of the water.

One of the attractive features of IGNP is its explicit objective of enabling small-scale and marginal farmers to take advantage of the irrigation, giving them preference in the settlement of the area. Priorities in the selection of the settlers are clearly defined in the rules for settlement (regularized in 1972). The landless agricultural labourers and the small-scale farmers residing closest to the villages to be settled have first priority, followed by similar classes of people from the same *tehsil* (administrative unit above the village), district, or adjacent district.

Until 1980, practically all allocations went to households in these categories. Among these, priority went to households of the scheduled castes and scheduled tribes. Each settler received about 25 *bighas* (6.32 ha) of land. These households bought the land at a reserve price, which was considerably lower than the market price, and were given 15 years to pay off the cost. Only since 1988 (since 1980, in stage I) has 50% of the unoccupied government land in the command area been reserved, at market price, for people in nonpriority groups. However, this failed to alter the composition of the settlements in the area. As of June 1991, out of 25 678 allotments in the stage II area, only 743 went to people in nonpriority groups (Hooja 1994). One cannot find fault, therefore, with the targeting of the allotments. However, the outcome was less than desired. The pace of settlement was also slow. A number of settlers never took possession. Many new settlers had to hypothecate their lands or became sharecroppers, as they were unable to earn enough from their holdings. Also, the phenomenon of absentee landholders distorted the outcome.

Several factors explain the poor rate of settlement and unsatisfactory increase in the income of the settlers. In the first place, not all of the area allotted to the settlers was irrigable. Of the standard allotment of 6.32 ha, less than one-half in the stage II area (only 46%) was under the command area of IGNP. Besides, the potentially irrigable land could only be partially irrigated because of the difficult terrain (undulating lands, faulty construction of watercourses, and large sand dunes), which required high investments in land leveling and desilting of watercourses.

A number of allottees never settled on their allotted land but retained their rights to it so as to accumulate its asset value. This is particularly true of those few original inhabitants of the area who had permanent or quasipermanent ownership rights to the land before it came under the command of the project. Their major occupation continued to be animal husbandry (Mathur and Gurjar 1991). Another group of people who hardly occupied the land allotted to them were the *onstees* from Himachal Pradesh, who had been given land in the IGNP area when they were displaced by the construction of the Pong dam on the Indus River. For a variety of reasons these people cultivated very little of their allotment, although several turned their land over to local people or other settlers on a crop-sharing basis.

Also, many people who originally came to cultivate the land in the stage II area left after a few years, as the gestation period for making the land suitable for a reasonable level of agriculture proved to be very long. The paucity of settlers on the land, in turn, made it more difficult to develop the area. For example, wherever cultivated holdings were noncontiguous, the field channels were impossible to desilt, making it more difficult for the few remaining settlers. Similarly, without an adequate number of clients, supportive agencies — such as the banks, input depots, marketing depots, and educational and health institutions — were reluctant to start operations in these sparsely populated areas.

A team from the Indian Institute of Management, Ahmedabad (IIMA), conducted an in-depth study of the reasons for the slow movement of settlers into this area (Seetharaman et al. n.d.). Major problems identified by the IIMA team were uncertainties about the availability of water; unavailability of tractors at reasonable charges for leveling and tilling; unavailability of credit for agricultural inputs; and, in a few *chaks* (blocks of holdings below the village, or *abadi*, level), unavailability of drinking water. After trying to relate progress in the settlement to some basic determinants, the team concluded that

+ The extent of fragmentation of land seemed to affect settlement — the greater the fragmentation, the lower was the rate of settlement;

+ The distance of a *chak* from the nearest *diggi* (drinking-water reservoir) affected the settlement rate — the greater the distance, the lower was the rate of settlement;

+ The distance from the nearest *abadi* (bigger village) appeared to have no significant impact on the rate of settlement;

+ The distance from the nearest road had no significant impact on the rate of settlement;

+ The distance from the *mandi* (marketplace) had a significant impact on the rate of settlement — the greater the distance, the lower was the rate of settlement; and

+ The dominance of a single caste in a *chak* appeared to have little bearing on the rate of settlement.

As a result of the initial difficulties in cultivation, a class of absentee landowners emerged. These original allottees leased their land to poor farmers on a sharecropping basis. Sharecropping in the IGNP area is estimated to take place on 30–50% of the allotted lands (Ramanathan and Rathore 1991). Because of credit requirements, a large number of poorer cultivators in the initial phase had to borrow from the more affluent farmers and enter into exploitative labour contacts with their creditors (Sharma and Rathore 1990).

The World Food Programme (WFP) provided, under its Project 2000, assistance to the settlers during their initial period of settlement to enable them to continue on the land. Food was provided to new settlers for 24 months from their date of arrival or from the time when adequate water was made available if it had been unavailable when they arrived. By the end of this period, the settler was expected to be harvesting the fourth crop and to have attained a level of production sufficient to cover basic needs. Each settler's family was also granted an interest-free loan of 2 000 INR (which has since been increased to 5 000 INR). An evaluation of the project indicated that WFP assistance, although most welcome, provided only a small part of the credit needed for such costly operations as land leveling. Also, the number of beneficiaries and the actual amounts they received were lower than the targeted figures (Singh 1994).

Two other handicaps plagued the settlers. First, most of the allottees came with hardly any background in irrigated farming. In fact, many were landless agricultural labourers — in a few cases, nomadic pasturalists. The agricultural-extension resources were inadequate to cope with the problems of these new settlers. Second, no thought was given to ensuring that settlements had people with similar social and cultural backgrounds. The settlers were selected for lands by drawing lots, which could result in total strangers being made to cultivate land as neighbours. However, in recent years, more thought has been given to selecting and settling allottees in cohesive groups.

In an insightful paper, Mathur (1991) summed up the preconditions for a successful settlement process in a hostile environment:

> The task [of the settlement of the poor] in this difficult region can be successfully accomplished only if: (a) people possess enough material resources to invest in agriculture; (b) people know the technique of irrigated farming; and (c) the people are prepared to live and work together in a difficult, new environment.

In IGNP, these preconditions were not met. Planners expected either the market process or administrative intervention to ease the situation of the settlers. Unfortunately, this never took place.

Market and institutional failures

To understand the issues and problems specific to the stage II area, the IGNB and the office of the Commissioner for Area Development sponsored a host of studies. On the basis of these studies and their understanding, it appeared that the main needs were the following:

+ Organizational coordination among different agencies functioning in the IGNP area and strong motivation for government functionaries at various levels to act;

+ Meaningful participation of the settlers in management of the project;

+ Generation and extension of relevant hydrological and agronomic knowledge; and

+ Creation and maintenance of basic infrastructure.

The markets and institutions in the region were thought to be one means to address these challenges. But markets failed to perform the job of allocating resources and rewards in support of the project's objectives. Similarly, the institutional backup was weak at different levels, so bureaucratic and nonbureaucratic players did not receive the correct guidance or signals.

Four interrelated factors may have contributed to the failure of the market in meeting the project's objectives: high transaction costs; interlocked markets; lack of information and the high cost of information that was available; and weak demand. Initially, sparse population and low household incomes limited the size of the market for goods and services. The absence of supportive services, such as marketing and credit, was a result of the limited market. Because of the small number of clients, the unit cost of delivery was, naturally, very high. The limitation imposed by the size of the market was compounded by the lack of infrastructure, especially roads. Limited access resulted in thwarted competition and rising costs of goods and services. The net result was that inputs required for agricultural development were never delivered to the settlers, and the farmers had no access to cheaper credit through formal institutions.

Access to timely and adequate credit was a critical factor given the initial poverty of settlers and the long gestation period they needed to make their land productive in such a harsh environment. Because formal sources of credit were inaccessible, informal sources of credit, such as traders, commission agents, and large-scale farmers, had to be used. This led to the interlocking of the credit market with the trade market if the credit came from traders and the commission agents; with the labour market, if credit came from large-scale farmers. As a result, the settlers did not receive proper rewards for their products and labour. Allottees saw this as a disincentive to settle on the land allotted to them.

An important cultural factor was the low level of literacy and, in turn, limited access to information, which created another set of handicaps. Information generally leads to entitlement, and ignorance generally leads to deprivation. Neither the official nor the unofficial agencies succeeded in creating an awareness among the settlers of the importance of literacy as a tool for empowerment. The small coterie of people who had access to information, including the market functionaries, were able to take full advantage of the situation. When clients have imperfect information, markets for goods and services become more monopolistic.

Poverty, ignorance, and a sense of dependence weakened the farmers' influence over the research and extension system, the input delivery system, the canal system, and the system of development priorities. The settlers had a weak voice and hence no position of influence (Paul 1992). In other words, the resource-poor settlers could make the research, extension, or input delivery systems demand driven, and, to that extent, their market influence was marginal. Because of the overall situation in the stage II area of IGNP, the markets alone where unable to allocate resources and generate incomes to support the project's broad objectives. In such circumstances, institutional interventions are needed to address the distortions induced by the market's failures. Unfortunately, the agencies involved in the settlements were unable to correct the market distortions. If anything, they exacerbated existing anomalies. The institutional failures occurred in the functioning of the bureaucracy, as well as in the organization of the people at the grass roots.

Government agencies failed to meet the challenge for several reasons. First, bureaucrats, in their role as agents of change, suffer from several handicaps. In this case, the bureaucrats' outlook was myopic, and their interaction with other important agents of change (for example, researchers, social activists, politicians) was minimal. When apparent, it was guided by no shared concerns or identified mutual interests. As resources are mainly controlled by the bureaucrats (in this type of development strategy), such behaviour marginalizes other sectors of the population, who lose the initiative to contribute to the growth of the economy or to society. Second, in government agencies, power is centralized in a hierarchical structure, and any initiative shown by functionaries at lower levels is viewed as suspect. As a consequence, bureaucrats on the lower rung hesitate to look beyond the implementation of orders received from above (Korten and Siy 1988). Third, bureaucrats lack accountability. Audits exist to ensure that inputs are used in the prescribed form and manner. However, hardly any system of accountability exists for the outcome of the efforts (Paul 1992). In bureaucracy-led development, it becomes extremely difficult to decide who is responsible for a decision — an individual or a group of individuals. Finally, bureaucrats have no long-term vision to guide their actions. At best, they may have far-sighted political masters. But if their political masters are short-sighted too, the bureaucrats have nothing to guide their actions to achieve long-term objectives.

With the IGNP, all these handicaps were compounded by the harshness of the terrain and the extremely difficult living conditions. These conditions weakened the line of control from the central bureau to the local functionaries. The functionaries should have lived and worked on the site but had all the reasons for not doing so and, because of extraneous considerations, could not be disciplined or motivated to deliver goods to the settlers. Added to these various handicaps, frequent transfers of officials made the situation much more difficult. Together, these weaknesses impeded the bureaucracy's functioning as an effective development agency.

Along with effective bureaucratic accountability and quality control, local initiative was conspicuously absent. Several factors explained people's lack of involvement (Srivastava and Rathore 1992). Heterogeneity of the groups made cooperation difficult. The poverty of the people made them heavily discount future incomes and therefore discouraged their taking long-term development measures on their own. Absence of genuine people's organizations at the local level worsened the situation. Although formal *chak samities* (committees of the beneficiaries at the *chak* level) were set up in various parts of the region, they made hardly any contribution. Above all, the approach taken to development tasks created a dependency syndrome.

Concluding observations

The economic position of the settlers will remain weak and their contribution to ecological sustainability, as well as agricultural development, will remain suboptimal unless some measures are taken to strengthen the market forces and make bureaucratic interventions more effective and people's involvement genuine and extensive. Two prerequisites for successful operation of the markets in this region are the creation of necessary infrastructure and the far-reaching dissemination of information. Even in areas where these two conditions are met and such areas expand over time, effective market functioning cannot be taken for granted.

To overcome the limitations of markets in such circumstances, Robinson and Tinker (this volume) advanced the concept of constrained markets. They suggested two types of constraints in this context (p. 38): "boundary, or external, constraints that set limits within which the market may operate; and target, or internal, constraints that alter the market value of goods, services, and resources." It is important that even with such constraints on the markets, the recipient system — the settlers — should have sufficient bargaining power to enter into transactions with the delivery systems on equal footing. The cooperative form of organization at the grass roots could probably impart such strength to the settlers. However, we know enough about the preconditions for the formation and successful functioning of such groups to be swayed by the number of formally registered cooperatives in India: small groups of households with similar social and economic backgrounds have been able to increase the bargaining power of small producers. Some form of collective action by the settlers in buying and selling goods and services would be a precondition for proper functioning of the markets in the IGNP areas.

For bureaucracy to be effective, it is most important to strengthen the last link in the chain — the state functionaries at the *chak*, or village, level. The government should give serious attention to motivating people at that level, both through financial incentives and nonfinancial benefits, together with effective deterrents for lapses of duty. However, a precondition for effective delivery of services to the settlers is that these functionaries reside or settle in the midst of the people they are supposed to serve. At the moment, all government services expected at the local level are provided by remote control. The functionaries live in nearby or not-so-nearby towns or cities. Unless the settlements have facilities for minimum levels of education, health, and housing, it is unrealistic to expect the functionaries to settle there. Government services could be organized in a hierarchical mode, with planning and coordination at the district level and delivery at the grass-roots level. Appropriate facilities would be available to the functionaries at different levels.

The three government agencies directly involved in the development of the IGNP area — the IGNB, the Commissioner for Area Development, and the Commissioner for Colonization Development — should have much better coordination. Coordination is equally important at the local level. The bureaucrats should become sensitive to the social and ecological aspects of development. A coalition with the nongovernmental organizations (NGOs) functioning in the area would also be helpful. This coalition of the government organizations and NGOs should also involve the researchers working in hydrology, agronomy, and the social sciences. At least an informal forum for periodic exchange of views on important development issues is needed. It is equally important to ensure follow-up of the decisions made at such a forum.

In addition to supporting people's initiative and involvement in organizing small, homogeneous collectives for economic action, it is important to increase the political clout of the settlers. This should be possible with the strengthened Panchayati Raj institutions (local self-government institutions at the village, *tehsil*, and district levels), which were given statutory powers and resources through recently amended constitutional provisions. More responsibility should be devolved to these institutions. An important step would be to give the responsibility for supervision of local-level functionaries to the elected representatives of the people at the corresponding level. Subsidiarity would be an innovative way to ensure the advantages of both local initiatives and cooperation within the overall system. Subsidiarity is the delegation of managerial powers to local units in return for legally binding agreements on basic environmental rights and responsibilities at the household and community levels (Robinson and Tinker, this volume).

The results of devolving authority for development tasks to the Panchayat level have been very encouraging. West Bengal is a good example, with its active Panchayats at the village level. Karnataka, *was* another, during the brief time when a genuine Panchayati Raj system functioned there.

The development tasks in the IGNP area are difficult, but the possibilities of meeting some of the important objectives of development and contributing to sustainable growth are also very real. The markets, bureaucracy, and people's cooperatives will have a distinct role in achieving these goals.

Chapter 9

Integrated Research and Policies in East Asia

P.S. Intal, Jr

⋄ ✦ ⋄

Development and environment need not be strange bedfellows; the challenge is to redirect the pattern of development, as well as the policy and institutional environments, to more effectively manage the trade-offs among economic growth, social equity, and environmental protection.

Immiserizing growth and resource degradation in developing countries

Resource depletion and environmental degradation are not new phenomena in the developing world. Much of the developing world opened up its trade with the West by cutting down forests and using river deltas and grasslands for planting crops or ranching cattle for export during the colonial era; this is the historical foundation of the so-called vent-for-surplus theory in international-trade literature. For example, the whole island of Cebu in the Philippines was stripped of its hardwood trees to build Spanish galleons during the 17th and 18th centuries. The difference between then and now is that the resource and environmental degradation in much of the developing world has shifted from being highly localized to being more widespread and pervasive, with attendant negative impact on the sustainability of agriculture, fishing, and forestry.

The sharp increase in population in recent decades is certainly an important factor in the depletion of resources in many developing countries. But equally important are the development and policy failures in many developing countries, which have immiserized growth for significant periods. During the 1950s and 1960s and, for some countries, even the 1970s, the promotion and protection of heavy industry led to the misallocation of limited investable resources in industries that had no comparative advantage. Per capita income barely improved with free-trade prices, resulting in boom–bust growth cycles (that is, immiserizing growth). Prolonged and indiscriminate industrial protection created an import-dependent, capital-intensive, and increasingly technologically obsolete manufacturing sector. Import dependent but not export competitive, the industrial sector became a net user of foreign exchange for too long.

As a result, continued pressure was put on the agriculture and forestry sectors to generate export earnings. Public subsidies were used so that agricultural products and livestock could be exported to raise foreign exchange, which accelerated the conversion of forest lands into agricultural lands and pastures (Brazil, for example). This was coupled with the rising social costs of resource extraction through resettlement schemes, the degradation of ecosystems supporting subsistence livelihoods, and the more general inequitable access to natural resources. Finally, the rent-seeking opportunities from natural-resource extraction in many parts of the developing world led the political and economic elite to appropriate for themselves access to commercially profitable natural resources, through licenses and permits, while resisting attempts to increase logging and franchising fees to more realistic levels.

Because of the growing population base and the low level of labour absorption in industry, the agricultural and rural sector, together with the informal urban service sector, has become the natural valve for employment. In many developing countries, like the Philippines, population growth and poor employment prospects in industry have led to increased pressure for further parcelization of cultivated land and a greater push toward the open-access, marginal, and public (usually deforested) uplands (Cruz et al. 1988; Cruz and Repetto 1992). The poor state of rural infrastructure, the pricing bias against agriculture, uncertain property rights, and poverty have been disincentives to the widespread adoption of productivity-enhancing and conservation-intensive agriculture and agroforestry.

In short, the resource and environmental degradation in many parts of the developing world in the past three decades reflects an immiserizing-growth trend. This involved heavy and prolonged industrial protection, rent-seeking and unequal access to natural resources, and direct and indirect taxation of agriculture and the natural-resource sector, all in the face of a growing population base.

Solutions to these problems are multifaceted and both procedurally and technically substantive. The developing world needs a more competitive and productive industrial sector, increased industrial employment, and net positive (instead of negative) resource transfers to the agricultural and rural sector. Natural-resource extraction should be priced to include the social costs of environmental damage and resource extraction beyond a sustainable level. Public investment should be made in human-capital formation, resource regeneration, and environmental improvement. With increased educational attainment and improved employment opportunities, especially among women, the concomitant higher and more equitable growth will improve the effectiveness of efforts to reduce population growth (see, for example, Mason 1997). These changes require economic openness and a competitive and environmentally friendly regulatory and institutional framework.

Growth and the environment in East Asia

In many respects, successful East Asian economies are the best examples of the global shift to economic openness and export orientation. The remarkable economic performance of East Asian countries since the 1960s is now well known. In the past few decades, the economic landscape of East Asia has been changed by the economic recovery and maturity of industrial Japan, the rapid economic transformation and industrialization of the Asian newly industrialized economies (NIEs; Hongkong, Singapore, South Korea, and Taiwan), and lately, the surge in growth and acceleration of structural changes in the ASEAN-4 (Association of Southeast Asian Nations; especially Indonesia, Malaysia, and Thailand) and China. The success stories of East Asia's dragons or tigers have led to a growing literature that attempts to explain such remarkable economic performance (see, for example, World Bank 1993; Amsden 1994; Ranis 1995; Akyuz et al. 1997).

Despite controversies in the interpretation of East Asian experience, especially of the role of the government vis-à-vis the market, the experience of the successful East Asian countries brings out, at a more fundamental level, the importance of hewing to the intuition of basic economics. Specifically, five conditions of the East Asian success can be highlighted (Intal 1995):

+ Economic outwardness and export push, which encourage the allocation of a country's limited resources in accordance with its evolving international competitiveness;

+ Macroeconomic and price stability, which minimizes uncertainty and the transaction costs of intertemporal decisions, thereby encouraging higher savings and investment rates;

+ General flexibility of the domestic factor markets, both capital and labour, which maximizes the economy's capacity to adjust to market shocks and shifts in international competitiveness and comparative advantage;

+ High investments in infrastructure and human-capital formation, thereby forging an integrated economy and increasing people's capacity to adjust to changes in market fundamentals and technology; and

✦ A tendency not to overly tax the agricultural sector and not to overly subsidize or protect the industrial sector to prevent greater inequality across social and income classes.

These conditions encourage efficient allocation of resources, greater focus on productivity, and acceleration of technological adaptation and upgrading. In addition, the governments of the successful East Asian economies propagated the principle of shared growth, implementing major equity-oriented programs to ensure that economic growth benefits everybody (Page 1994). Shared growth and strong government–private-sector linkages in policy decision-making helped ensure society's sense of "ownership" of the governments' economic programs. In some respects, the East Asian experience has brought a major revolution in development-policy thinking, turning sharply away from the Latin American-centred import-substitution development strategy of the 1930s through 1960s, toward the strong export-oriented policy bias of the 1990s.

This chapter highlights the surge in economic growth and acceleration in industrialization of the successful ASEAN-4 countries and China in recent years. With the exception of the Philippines, these countries have had rates of economic growth during the past decade or so among the highest in the world. They have become markedly outward oriented, as indicated by a significant rise in exports and imports relative to gross domestic product (GDP). These high economic growth rates have been accompanied by a sharp rise in investment and domestic savings rates, as well as by a significant reduction in agriculture and a significant increase in manufacturing as proportions of GDP. Despite the decline in agriculture as a proportion of GDP, the net export surplus (exports minus imports) from agriculture and natural-resource trade increased in Indonesia, Malaysia, and Thailand during the latter part of the 1980s. During the same period, China turned from being a net importer of agricultural and natural-resource products to being a net exporter.

The Philippines stands out as an exception to the successful performance of the ASEAN-4 countries during the 1980s and early 1990s. The reasons for this were the economic crisis of the mid-1980s and the difficult process of adjustment toward macroeconomic stability and structural reforms. However, in recent years (1995–97), the macroeconomic fundamentals in the Philippines point to a growth trajectory that is likely to mimic that of the rest of East Asia.

Whether from fast growth or from sluggish growth, East Asia's environment has been under a heavy strain in the past few decades. Much of the Philippine experience has followed the dynamics of immiserizing growth, resource depletion, and environmental degradation described for the developing world earlier. For the fast-growing Northeast Asian countries, the relationship between environment and growth follows an inverted-U pattern in relationship to per capita income; that is, many pollutant levels tend to increase with rising per capita incomes but level off and finally decline at substantially higher per capita incomes. Different pollutants have different per capita-income levels for leveling off and eventually declining. The inverted-U pattern arises from the conflicting effects of scale of production, shifts in demand and pattern of production, technological change, consumer demands for environmental amenities, and the capability of the government to enforce environmental regulations (O'Connor 1994).

In Japan, the shift in manufacturing production from the less pollutive food and light manufacturing industries in the mid-1950s to the more pollutive chemical, oil, pulp and paper, and primary-metal industries in the 1960s brought such infamous incidents of pollution as the Minamata mercury poisoning. By the 1980s, the pattern of manufacturing in Japan shifted to the less pollutive (in terms of toxic waste) industries like electronics and electrical manufacturing; moreover, the proportion of manufacturing in GDP also declined as the Japanese economy matured.

The same pattern of rising then falling pollution intensity of manufacturing production is evident in Taiwan, reflecting in part the changing composition of manufacturing production, from food, beverages, and textiles and garments in the mid-1950s to the more pollutive industries, such as chemicals and petroleum and steel, in the 1970s, and thence to the less pollutive electronics and electrical industries in the 1980s and early 1990s (O'Connor 1994).

There are indications of increased environmental degradation in the ASEAN-4 countries. Particulate-matter levels, for example, are higher in the capital cities than the ambient standards. Thailand has turned from being a net exporter of forestry products to being a net importer, a reflection of the fast denudation of its forest reserves. The Philippines is nearly in the same situation with respect to forestry resources and trade. Forest denudation was also considerable in Indonesia and to a lesser extent in Malaysia during the 1980s. Both Thailand and the Philippines have substantially increased their fishery exports, primarily as a result of the conversion of mangroves into fishponds, which substantially reduced their stocks of mangrove forests. Environmental problems in Thailand's major cities are worsening, including air pollution from increased levels of industrial emissions (apart from vehicle emissions), water pollution from untreated waste water of industries and households, and hazardous and toxic wastes from industries and hospital and household refuse (Phantumvanit et al. 1994). The environmental problems of the major cities in the Philippines are also worsening, not only because of industrial wastes but also, and more importantly, because of household wastes. Indonesia's major cities are probably facing similar problems. Even in the less populated and more prosperous Malaysia, air pollution is approaching critical levels in urban areas like Kuala Lumpur, and water pollution and land degradation continue to be major concerns (Ali et al. 1995).

The historical experience of the developing world and of East Asia indicates that the environment has been under a heavy strain from a persistent lack of or a lopsided type of development (as has been the experience of the developing world during much of the post World War II period) or from rapid economic growth. The inverted-U pattern of pollution levels in relation to levels of per capita income gives some indication of the reversibility of many sources of pollution, which means that development and economic growth are needed for countries to attain levels of per capita income that would put them in a trajectory of declining pollution. The key issue is whether the magnitude of the trade-off between the environment and economic growth can be reduced at each point in the development process.

The challenge is to reduce the amplitude of the inverted-U relationship such that at each level of per capita income the level of pollution is lower. In the process, the cumulative magnitude of environmental degradation would be lower than in historical experience. In developed countries, product and process innovations are increasingly directed to greater energy efficiency, higher material recovery, and lower waste generation. Evidence suggests that developing countries (as industrial latecomers) can draw from the ongoing technological developments in developed countries to reduce the trade-off between economic growth and the environment. The challenge is to create the policy and institutional environment and mechanisms to bring about greater capital formation (machines with better environmental attributes), stronger incentives for resource regeneration and environmental protection, and more robust and sustainable growth. These are the fundamental problems facing integrated research and the formulation and implementation of integrated policies.

Policies and the environment: a framework for integration

The environment cannot be protected and conserved by environmental management alone (Tan and Intal 1992). Environmental management without harmonization with economic and demographic policies is inefficient and costly. Similarly, economic policies without consideration of environmental consequences lead to resource depletion and environmental degradation. Such policies, even if successful in the short term, will ultimately fail because of the lack of, or serious deterioration of, the resource base.

Figure 1 presents a framework for understanding the link between policy-making and the environment. In economic planning and policy-making, policy instruments are directed either to the economy as a whole (for example, macroeconomic policies) or to a specific sector (sector-specific policies). The impact of government policies on the environment can be direct or indirect. Direct impact comes from policies whose principal objective is to affect the environment, such as a reforestation policy. The indirect impact on the environment arises from the externality effects of production associated with a given policy. Policies also affect the incidence and magnitude of poverty, which is acknowledged as being a key contributor to resource depletion and environmental degradation.

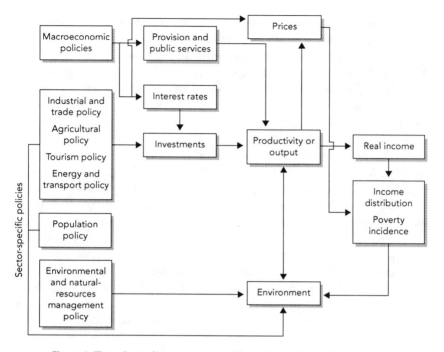

Figure 1. The policy-making–environment linkage. Source: Tan and Intal (1992).

The impact of macroeconomic policies (for example, monetary policies, exchange-rate policies) on the environment is indirect, primarily via their effects on industry-sector output levels. The increased output of a natural-resource-intensive or highly waste-generating industry places increased stress on the environment. In contrast, environmental-management and natural-resource-management policies are the most direct tools the government can use to have an impact on the environment. These policies may take the form of direct intervention in the environment and the natural-resource sector, for example, through reforestation and licencing. Policies may include economic incentives to encourage concern for the environment and the efficient use of natural resources; the assignment of unambiguous property rights and the correct valuation of natural resources would fall into this category.

Industrial, agricultural, energy, trade, and population policies also have significant impacts on the environment. Their impacts include the pollution effects of sectoral production and also come through the poverty and income distribution channel. The impacts of population policies on the state of natural resources and the environment come through such channels as food requirements, migration (at times to environmentally fragile areas), and waste management. Sustainable development therefore hinges on the capability of the decision-makers to consider both the direct and the indirect impacts of policies on the environment and to balance them with economic-growth objectives, that is, by designing and implementing integrated policies.

Behind the environmental problems are economic-policy distortions, institutional inadequacies, and sociopolitical disparities. Sustainable development is really much more about economic, institutional, and political reforms than about the environment per se. The concern for the environment forces us to confront the economic, institutional, and sociopolitical failures that have led to its serious deterioration.

[The drive for environmental protection can encourage governments to adopt better economic policies.]An example of this is an agricultural policy that reduces tariffs on corn, discourages upland corn production (which erodes the soil), and encourages livestock production in the uplands, especially in conjunction with investments in fruits and agroforestry, which are less erosive alternatives. On the other hand, water pollution from livestock waste can be a problem if livestock production is intensive and the upland area is a watershed for domestic water sources downstream. However, reduced tariffs improve the overall allocation of resources within the agricultural sector. With better pricing policies, moreover, input subsidies can be reduced or at least better targeted, which, by itself, contributes to better use of resources. Thus, economic policies informed by environmental considerations are better designed and more socially beneficial.

The drive for environmental protection also encourages institutional innovations that have beneficial economic effects and improve governance. Here, the Philippine experience with industrial environmental management may be relevant. A study of the Philippines under a United Nations Conference on Trade and Development project (Intal et al. 1994) indicated that the proportion of incremental abatement costs in the value of total industry output is relatively small. In interviews, many chief executive officers of top corporations in the Philippines said that stronger implementation of environmental rules would not jeopardize their international competitiveness. Equally important, a number of cases in the Philippines suggest that waste-minimization investments result in cost savings. Pricing policies that internalize environmental externalities encourage waste-minimization practices and improve the internal operating efficiencies of firms.

On the institutional side, government has the potential to improve its approach to industrial environmental management. At present, the bureaucracy in the Philippines is unable to meet the demands of increased environmental protection. Lack of personnel and inadequate skills bedevil the environmental-impact assessment (EIA) process, especially for major-investment projects. Delays in the process increase transaction costs for private investors. Monitoring of environmental violations suffers from a lack of laboratory facilities, especially in the regions, and relies on complaints from affected communities. Local officials have used environmental regulations to harass firms, especially during elections. The Philippine government is modifying its approach to industrial environmental management, away from a centralized command-and-control (CAC) approach to one that

+ Uses programmatic EIAs, especially for industrial estates, to reduce transaction costs of EIAs for private investors;

+ Encourages firms to provide pollution-management appraisals (PMAs) and adopt waste-minimization approaches (PMAs are likely to be consistent with ISO 14000);

+ Uses private laboratories and nongovernmental organizations (NGOs) in the regions to monitor industrial pollution and violations; and

+ Develops market-based instruments, in addition to CAC.

Explicitly integrating environmental considerations into policy-making can both influence the substantive approach to economic policies and help shape the nature of institutional and procedural changes needed to improve governance.

Integrated research and policy: examples from the Philippines

Research has three broad roles as far as policymakers are concerned (Glover 1994). The most important and enduring is setting the terms of the policy debate, that is, presenting the perspective from which an issue or phenomenon is viewed by the public and public policymakers. In this role, cumulative research helps set the framework for discussing and evaluating problems and issues. The second role of research is evaluating ongoing or past decisions and developments to determine lessons that can be applied to improve current decisions and programs. The third role is evaluating developments, concerns, and opportunities as they emerge. The first role focuses on setting the policy direction, and the second and third roles help policymakers determine or refine a specific set of policy actions or programs. Research on the impact of social-science research on policy-making in the United States suggests that research in the first role has the most enduring impact, whereas research in the third role has been ineffective.

An inadequate supply of competent policy-comprehending researchers and of research funds plagues the research community in developing East Asia, especially in the fields of natural resources and environment. In addition, developing East Asian countries have insufficient financial resources to make investments in resource and environment regeneration and protection. As a result, the researchers strongly interact with the policymakers; at times, the researchers themselves become policymakers. Also, the donor community has significantly influenced the research and policy agenda. The linkages among key researchers, policymakers, and the donor community have shaped the research process and the role of research in policy-making and program design.

Examples from the Philippines provide some insights into the role of research and its impact on policy-making, especially in the fields of natural resources and environment. The examples show the usefulness of bringing social-science input into natural-resource and environmental policy-making, as well as of bringing environmental considerations into socioeconomic policy-making. The examples also indicate the importance of a strong interaction, including personal interaction, between researchers and policymakers. Contrary to experience in the United States (Glover 1994), researchers have had a greater impact on policy-making in developing East Asia. This is due, in part, to the research and policy elite in these countries being fewer than in the United States and, in part, to the more frequent interaction among key researchers, policymakers, and the donor community with respect to natural-resource and environmental policy reforms and program design. The following subsections provide examples from the Philippines: the social forestry programs; a mangrove valuation study; and an environmental and natural-resources accounting project.

Social forestry programs

The social forestry programs of the Philippine government are the offshoot of two initially parallel but ultimately interdependent efforts. The first is the effort of government to develop programs to tackle the problem of deforestation, brought about by the upland population, and thereby help rehabilitate forest lands, stabilize forest occupancy, and improve socioeconomic conditions for upland dwellers. The second is the effort of social and physical scientists to understand the nature, causes, and impacts of forest-management systems, as well as the impacts of government programs on the uplands.

From the Spanish colonial period, in the 1800s, up to the mid-1960s, the official government policy was to prohibit occupation of forest reserves, watersheds, and national parks; any occupants were threatened with legal prosecution and forcible ejection (Tolentino 1991). For example, in 1964, investigators won convictions against 136 *kaingineros* (shifting cultivators) and investigated 655 cases involving *kaingineros* (NCKP 1964). By 1975, the Philippine government had officially acknowledged bona fide forest occupants of the forest lands, including pioneer migrants, who were of cultural minorities. Rather than ejecting and resettling the occupants, the government introduced programs for *in situ* forest-protection activities. Finally, by the late 1980s, the upland settlers had become a critical foundation for the government's strategy for forest renewal and protection.

Programs for forest protection and renewal involving upland dwellers started with the Forest Occupancy Management (FOM) program, in 1976. The program granted upland dwellers 2-year forest-occupancy permits for a maximum of 7 ha. This was the beginning of the government's *in situ* programs for improving the socioeconomic conditions of upland dwellers while pursuing the objectives of forest rehabilitation and protection. FOM was followed, in 1979, by the Family Approach to Reforestation (FAR): families were hired for 2 years, on a contractual basis, to do reforestation. Also in 1979, the government instituted the Community Tree Farming (CTF) program, granting CTF leases for 25 years (renewable), with a maximum of 2 ha for individual parcels for low-income farmers.

In 1982, the Integrated Social Forestry Program (ISFP) integrated all these projects, granting individual and communal stewardship contracts for 25 years (renewable) for at most 7 ha per individual. Under ISFP, beneficiaries have the exclusive right to occupy and manage an ISFP plot of land. The government instituted the Community Forestry Program in 1989 and the National Forestry Program in 1990, and both were similar to ISFP. They involved 25-year forest-management agreements (renewable once) and allowed the parties to manage certain sites and to harvest, sell, and use forest products from these sites (de los Angeles and Bennagen 1993). The effective 50-year limit on leases conforms to the Constitutional limit for leases of land and natural resources.

Research and researchers on the uplands social forestry programs played a significant role in the remarkable shift, within two decades, in the official thinking on upland settlers. The introduction of the *in situ* forest-protection activities involving upland dwellers reflected the Bureau of Forest Development's realization that punitive measures like ejection do not work. The government's decision to grant 25-year contracts was influenced by research results and researchers' recommendations for longer (and hence more secure) land tenure than available with the 2-year leases initially granted by government.

One of the earliest examples of interdisciplinary research in the Philippines illustrates that at times, the most enduring impact of research on policy may differ from what was expected when the main concern of the research was defined. The University of the Philippines at Los Baños (UPLB) established its Upland Hydroecology Program (UHP) in 1976, with 20 scientists from various disciplines and units of the university. The objective of the research (Sajise 1984, pp. 313–314) was to

> broaden the basic understanding of upland ecosystem dynamics in order that schemes for land-use classification/allocation of upland areas and the management of swidden fields, grasslands, second growth forests, and industrial plantations as protective watersheds in the Philippines might be improved or developed appropriately.

The team consisted initially of natural scientists (agroeconomist, entomologist, soil scientist, meteorologist, chemist, animal scientists, taxonomist, etc.) but later included an economist and a sociologist.

The project initially focused on agroforestry technology, but the group wondered why the farmers in their project areas were not adopting the technologies (such as contour planting). The social scientists on the team discovered

the importance of the security of land ownership, labour availability and allocation, and the market in upland farmers' decisions about crop and production
technologies. As the team leader pointed out in an interview, if the program has
not been interdisciplinary, the team would have missed the tenurial aspects,
which turned out to have the most important impact on policy (Sajise, personal
communication, 1995[1]).

It is also worth noting that the UPLB–UHP team was not the first to recognize the importance of security of land tenure. For example, anthropologists
studying cultural minorities and upland dwellers in the Philippines had noted
the settlers' intimate knowledge of upland ecosystems and shown that the lack
of security of land tenure constrained their adoption of less ecologically destructive farming methods (Bennagen 1978). However, the emphasis that the
UPLB–UHP natural scientists put on land tenure resonated with their former
colleagues in key policy-making posts at the Ministry of Natural Resources.
Sajise emphasized that the team members' personal discussions and interaction
with key ministry officials played an important role in convincing the government to change its policy perspective and implementation strategies.

Another catalyst in the design and implementation of the social forestry
programs was the Central Visayas Regional Project (CVRP), funded by the
World Bank. Cebu (in Central Visayas) has one of the highest rates of share ten-
ancy in the uplands. Because of this, one of the consultants for the project suggested that the government give the upland farmers 25-year leases (called
stewardship contracts) to address their apprehensions (de los Angeles, personal
communication, 1995[2]).

The linkage between the researchers and the donor community must also
be emphasized. The Ford Foundation initially supported UPLB–UHP with the
objective of furthering social forestry through research, drawing on its initial
work in India. UPLB–UHP fed into the design of the much larger and better
funded United States Agency for International Development (USAID) Rainfed
Resources Project with the Ministry of Natural Resources, in the early 1980s.
The ample aid funds under the USAID project encouraged some recalcitrant
officials in the ministry to change their minds and provide more support for
social forestry and for the improvement of tenurial conditions in the uplands.
The initial programs — FOM, FAR, and CTF — were eventually upgraded and
integrated into ISFP. Under ISFP, as mentioned above, individual forest occupants are provided with renewable 25-year certificates of stewardship and
upland communities can also sign 25-year community forest-stewardship agreements. ISFP became important during the early 1980s, partly for political
reasons: national elections were being held and the stewardship certificates were
a good tool for generating public support for the administration's candidates

---

[1] P. Sajise, Head, SEARCA, personal communication, 1995.
[2] M. de los Angeles, Research Fellow, Philippine Institute for Development Studies, Metro
Manila, Philippines, personal communication, 1995.

(Sajise, personal communication, 1995[1]).

The implementation of the social forestry programs made it possible to use action-oriented research (generally involving interdisciplinary teams) to test community-based and participatory approaches to uplands development. For example, a social forestry project that UPLB set up in a neighbouring province demonstrated the importance of confidence-building measures, of working with people rather than for them, of flexibility in project implementation, and of the roles of women in social forestry projects (Mariano 1986). A social forestry project in the Visayas, undertaken by Silliman University, indicated the need to redesign some of the development packages that the government was using to improve the socioeconomic conditions of the uplanders (Cadelina 1986). CVRP undertook a number of innovations, although the initial focus of the social forestry component of the project was on reforestation, rather than on agroforestry (Penalba et al. 1994). For example, CVRP's social forestry component was the first pilot project in the country on community-based management of residual forests. The interplay between the action-oriented research programs of academic institutions and the social forestry programs of government enriched the design of the government's later social forestry projects. Recent research indicates that indeed the uplanders with more secure land tenure tend to adopt less erosive techniques of production (Rola 1987).

The Aquino administration, installed after the February 1986 revolt against the Marcos government, brought a surge of reformist initiatives. With human-rights lawyers, environmentalists, and management graduates (instead of foresters) in control of the Department of Environment and Natural Resources (DENR) (formerly the Ministry of Natural Resources), the policy focus shifted to forestry protection, reforestation, and the welfare of the forest occupants, rather than that of the logging industry. Many timber concessions were canceled, and the ISFP was pushed to cover more area. Reforestation gained momentum. The department instituted community-based management agreements, as well as forest-land management agreements (for plantations), all for 25 years (and renewable), thereby imitating the ISFP approach.

The significant changes in the approach to forestry management and uplands development during the late 1980s resulted, in part, from the strong policy and program backup provided by researchers, both natural and social scientists, especially through the Uplands Working Group, under the office of one of the undersecretaries (or deputy ministers). The working group was to monitor and analyze the social forestry projects and recommend policy options. Many of the recommendations became DENR memoranda (Sajise, personal communication, 1995[1]). Also, some of the key members of the working group were social scientists, whose research focus, up until then, had been communal irrigation. Their research revealed the importance of participatory approaches and organizing the community. Their initial research on communal irrigation had also been funded by the Ford Foundation, at about the same time as the social forestry projects. The similarity in approach in social forestry and communal irrigation enabled DENR to tap the research insights and expertise of the researchers in both areas.

In sum, research on the uplands was often interdisciplinary in approach during the 1970s and 1980s and helped change the government's perspective on uplands development and forest protection and renewal. This change in perspective was supported at the policy and program level through the significant linkages among the researchers, the donor community, and the policymakers.

Mangrove valuation study

Mangroves play a significant role in coastal ecosystems, providing forestry- and fishery-based goods and services, primarily to coastal communities. In the Philippines, municipal fishers rely on open-access coastal fisheries and, like the upland farmers, belong to the poorest income classes. Fishers constitute about 5% of the total employed population (ADB 1990), whereas the upland population accounts for 7–14% of the total population (de los Angeles and Bennagen 1993).

The area of mangrove forests in the Philippines was estimated at 500 000 ha in 1918; this dropped to about 242 000 ha by 1970 and to about 142 000 ha by 1988 (EMB 1990). This represents a continuous depletion of the country's mangrove forests. The quality of mangrove areas has also deteriorated, and only minor areas of virgin forest remain (FAO 1982). The deterioration of mangrove forests and marine support systems, like coral reefs, has contributed to the decline in fishery harvests since the late 1970s. This, coupled with the rise in the number of fishers, has resulted in the catch-per-unit effort dropping rapidly during this time.

The decline in the mangrove areas is primarily due to conversion of mangroves into fishponds. The total area of brackish-water fishponds rose from 73 000 ha in 1950 to more than 210 000 ha in 1988. Despite the 1980 government ban on conversion of mangroves into fishponds, mangrove areas continued to decline by an average of 3 700 ha/year, which parallels the increase in the area of brackish water ponds, at 4 100 ha/year. The conversion of government lands to fishponds involved the granting of 10- or 25-year fishpond lease agreements, for an annual rental fee of 50 PHP/ha plus some nominal local taxes (in 1998, 37.9 Philippine pesos [PHP] = 1 US dollar). A widespread impression is that the mangrove areas are substantially undervalued. Fishponds on government lands are thought to be less productive than fishponds on private lands, and private and public investment in mangrove development has not been very substantial. This undervaluation has led to the depletion of the mangrove forests.

Under the Fisheries Sector Program, funded by the Asian Development Bank (ADB), the Philippine government agreed to undertake several policy initiatives to improve management of coastal and marine resources. Among these policy initiatives were the rationalization of existing regulations on fishpond licences and a proposed increase in licence fees. The mangrove valuation study attempted to estimate the social costs and benefits of converting mangroves to fishponds. The study assessed the proposal to raise the annual licence fee to 1 000 PHP/ha.

In contrast to the initial social forestry studies, the mangrove valuation study was shaped by a specific policy issue identified by policymakers and a donor. As a condition of the Fisheries Sector Program loan from ADB, the Philippine government was to estimate the economic rent from fishpond leases. The research team was able to convince the government that the research should not only estimate the economic rent (involving essentially economic analysis) but also be an interdisciplinary analysis of the value of mangroves. A multidisciplinary research team of economists, marine scientists, forestry experts, and aquaculture experts was formed to conduct the analysis.

The research team investigated both on-site and off-site mangrove products and services, drawing on old-growth, secondary-growth, and mangrove-plantation studies. The team estimated the market and household use of wood, fishery products, nipa palm, litter fall (as fertilizer and fish food), and spawning services of mangroves (especially for shrimps and milkfish). The team also estimated the returns from aquaculture but was unable to estimate the services of mangroves for folk medicine and erosion and buffer control during typhoons.

The team found that the market values of mangroves (for forestry and fishery products) are highest for old-growth mangroves, followed by mangrove plantations. The nonmarket values were also significant, coming primarily from spawning shrimps and milkfish. The team found that aquaculture operations would be able to support the increase in the annual licence fee (from 50 PHP/ha to 1 000 PHP/ha). More importantly, social net benefits would be derived from converting the less productive, low-income fishponds back to mangroves. The study results supported an increase in the licence fee, not only to enable the government to capture a larger share of the economic rent but also to encourage the inefficient aquaculturists to leave the industry and return the leased lands to the government for potential reversion to mangroves (de los Angeles, personal communication, 1995[3]).

The mangrove valuation study ended in 1996. In effect, the study results provide information to the disputants on the issue of raising the licence fee, that is, the fishpond operators (who naturally oppose the proposal) and the government. The government has made no decision, which may involve legislative amendments.

---

[3] M. de los Angeles, Research Fellow, Philippine Institute for Development Studies, Metro Manila, Philippines, personal communication, 1995.

The Environment and Natural Resources Accounting Project

The idea for the Environment and Natural Resources Accounting Project (ENRAP) came when the World Resources Institute (WRI) published and disseminated *Wasting Assets* (Repetto et al. 1989). DENR approached USAID's Manila office for possible donor funding for WRI to conduct a similar study in the Philippines. DENR wanted to use the project for policy-advocacy purposes, and both DENR and USAID wanted WRI to undertake the project. However, USAID's procedure calls for bidding. WRI had to back out because its charter forbids it to build on projects. A joint Filipino–American consulting consortium won the ENRAP bid. The project started in 1991 and was still under way at the time of writing. The project team included experts in forestry, fishery, soil science, geology, energy, engineering, and economics.

The first phase of ENRAP focused on changes in national income accounts resulting from depreciation or appreciation of forest land-based resources, particularly dipterocarp forests, plantation forests, pines forests, mangroves, and rattan. The first phase compared two alternative methods of valuation: the net-price method, used primarily by WRI, and the asset-valuation method. The two methods gave very different results, with the net-price method calling for an average of 6.6% negative adjustment in the country's net national product during the 1970s and 1980s and the asset-valuation approach calling for far less, a 0.22% negative adjustment during the same period (IRG–Edgevale Associates 1991). Asset valuation is the economically more appealing concept but is far more demanding in terms of data requirements than the net-price method.

The second phase of ENRAP expanded the sectoral coverage of the first phase to include depreciation of soil, mineral, and fishery resources; unpaid services for disposal of waste into air and water; direct nature services (for example, amenity values); environmental damage; and unmarketed production of upland agriculture and fuelwood. In effect, the second phase attempted to prepare a full environmental and natural-resources accounting, not merely sector-specific accounts. Therefore, the study included both sectoral and economy-wide analyses. This phase gave estimated ENRAP accounts for two selected regions in the country.

The major findings of ENRAP were the following (IRG–Edgevale Associates 1994):

+ Nonindustrial sectors are as important sources of pollution as the industrial sectors;

+ The cost of water-pollution control to the Philippine economy is considerably higher than that of air-pollution control;

+ The cost to reduce the present level of pollution by 90% across the board is higher than the benefits of pollution reduction, suggesting that sectoral prioritization is needed in pollution control to focus on the major sources of pollution and the places where the benefits outweigh the costs of pollution control;

+ Unmarketed natural-resource inputs to subsistence production are significant in upland agriculture and fuelwood gathering;

+ Depreciation is more of a problem for the renewable resources (forests and fisheries) than for nonrenewable ones (upland soils and mines); and

+ Off-site damage caused by resource extraction is considerable.

Initial analyses resulted in some interesting points, such as the following:

+ Achieving NIE status through much higher economic growth, but with the same 1988 technologies and pollution-control and policy regime vis-à-vis the environment, would bring considerable pollution. Achieving higher economic growth with far less pollution requires the adoption of less pollutive technologies and stronger policy support for environmental protection (Mendoza 1994).

+ ENRAP estimates of net present values of alternative land uses can be used to analyze alternative-land-use issues (Cabrido and Samar 1994).

+ ENRAP estimates indicate that the proportion of pollution-abatement costs in the total value of production is generally very small, suggesting that user charges for natural-resource use and pollution-emission charges are feasible policy instruments (Intal et al. 1994).

+ Trade liberalization (with attendant currency depreciation in real terms) would, other things being equal, more than proportionately increase the outputs of the more pollutive industries (including forestry, agricultural crops, and livestock) and thereby raise the overall pollutive content of domestic production. This suggests that the drive for trade liberalization and exchange-rate correction necessitates a correlative enforcement of environmental-protection measures, such as stumpage fees for logging and incentives for using less erosive cropping patterns and technologies (Intal et al. 1994).

The process of generating ENRAP accounts in the Philippines project is worth noting. The team used a process-oriented, network approach, owing to the large data requirements of the project (for example, data on pollution from industries, farms, households, and governments; physical damage to affected parties caused by pollution; cost of reducing the damage; and physical depletion of natural resources). To implement ENRAP, it was necessary to

+ Develop methodologies for generating environmental and natural-resource accounts;

+ Investigate resource- and sector-specific areas of concern;

+ Formulate a strategy for institutionalization of the ENRAP processes; and

+ Apply the accounts in national-level modeling of the environmental impacts of growth and development strategies.

Concerned government agencies participated at the policy level, through a policy steering committee, and at the technical level, through in-house workshops and regular meetings (de los Angeles and Peskin 1994).

This exercise provided information that could accentuate or temper initial policy biases. The activity was an offshoot of demands from policymakers, who had recognized its potential to contribute to policy-making. Indeed, environmental and natural-resource accounting is an important component of the Philippine Strategy for Sustainable Development. The credibility of the exercise lay in the rigour of the estimation process and the quality of the basic data. The multidisciplinary approach of the research team brought relevant sectoral expertise to the estimation process and the analyses. The research team made use of available local and international studies. The project revealed gaps in data, which provided the basis for improvement in the Philippine statistical system.

Although ENRAP is still ongoing, the initial results have already contributed to the policy debate. The results of the forest-resource accounting (phase one of the project) were used by both DENR and the Senate Committee on Environment and Natural Resources as inputs to the policy debates on the proposed total ban on logging. The estimates from the second phase of the project have provided policymakers with some indication of the negative externalities and magnitude of environmental damage caused by firms, households, and even the government. The estimates have made possible an analysis of the impact of macroeconomic policies on the environment. The findings of the project have been regularly presented before the Philippine Council for Sustainable Development.

The ENRAP estimates will be more useful after they are refined and updated, particularly given that policy debates on environmental protection are becoming more urgent. However, to maximize the benefits of an information base like ENRAP's, it has to be institutionalized in the country's statistical system. This would enable analysts to make more robust estimates and analyses for the policymakers. Perhaps more importantly, the availability of time-series data on economy–environment interactions would allow analysts to evaluate the benefits and costs of alternative environmental policies — data that are presently lacking in the Philippines.

It was too early to determine the extent of ENRAP's impacts on the country's policy-making. Nevertheless, to the extent that ENRAP's results clarify economy–environment interactions and provide a good indication of their order of magnitude, economic policies will probably be tempered by such information, and environmental policies will gather greater policy support. In this sense, the impact of ENRAP could be as fundamental as that of the social forestry research. ENRAP has increased the credibility and policy profile of the key Filipino researchers in the project. Several government agencies, the legislature, and NGOs in the Philippines have asked the researchers, especially the team leader, to shed light on several environmental and natural-resource issues. As analyses based on ENRAP's results and related activities continue and provide newer insights on policy issues, the potential policy impact of the ENRAP research is likely to increase.

Implications for integrated research and policies

Integrated research can make a difference to the policy-making process at two key levels of analysis. The first and probably most basic is the microlevel. Here the interaction among environmental, economic, and sociopolitical variables is best understood by considering a given set of ecosystems. The second is the macrolevel. Here the challenge is to disentangle the varying direct and indirect, as well as major and minor, interrelationships among these variables for a region or country. Both levels of research are needed to sharpen policymakers' understanding of these interactions so that they can develop integrated policies. The key role of integrated research is to provide policy and institutional insights drawn from interdisciplinary microlevel studies of a given ecosystem and from macrolevel studies of environment–economy interactions.

Microlevel studies of ecosystems are in principle system wide and probably more multidisciplinary than macrolevel studies, as the former highlight the locational dimension of socioeconomic activities. Explicit analysis of an ecosystem calls for natural science, whereas understanding the dynamics of human interactions calls for social science. As UPLB–UHP showed, interdisciplinary research can provide unexpected surprises, with significant policy implications. According to Sajise (1984, p. 321),

> perhaps the greatest benefit generated by the UPLB UHP was the demonstration to traditionally and strongly disciplinary-oriented scientists that a greater understanding of the upland ecosystem can be attained if viewed in the context of contributions from both the natural and social sciences.

Similarly, the multidisciplinary team of the mangrove valuation study was able to provide a more thorough evaluation of the contributions of mangroves, not only in terms of forestry products but also in terms of important functions, such as shrimp and milkfish spawning. As a result, the government has a stronger basis for making policy decisions regarding licence fees for fishponds, as well as for making public investments in mangrove development.

The degree of multidisciplinarity in microlevel research depends on the phenomenon or policy question being analyzed. Indeed, some of the complementary research at the microlevel may even be disciplinary. For example, given that security of tenure is very important to the adoption of more ecologically sound agroforestry technologies in the uplands, a significant correlative research and policy issue is whether the 25-year lease, with a certificate of stewardship, is better than straightforward private ownership of upland areas. It is known that land titles are better than certificates of stewardship as collateral for a bank loan; therefore, it seems more probable that if upland farmers have private ownership of their lands, they would invest more in reforestation and agroforestry technologies. Likewise, attention is needed on the issues of ancestral land claims and encroachment into the ancestral lands of the country's cultural minorities. These kinds of question require social-science perspectives; hence, research on them would more likely be disciplinary, but it would also complement multidisciplinary studies on the uplands.

Both interdisciplinary and disciplinary studies are useful in disentangling the various channels of interaction among the economic, environmental, and social variables at the macrolevel. Because the interactions are complex, the knowledge base generated would likely be the result of cumulative research on various facets of the complex whole. Nevertheless, as the experience of ENRAP indicates, even a rough understanding of the interrelationships of the economy and the environment (derived with simple tools of analysis and an extensive database on the pollution externalities, environmental damage, and pollution-control costs) can provide a fresh perspective in mainstream policy discussions. For example, given that some industries are more pollutive than others, various trade-policy options and economic-growth paths will have different implications for the levels of pollution in the country. Similarly, given that the proportion of pollution-control costs in the total value of industrial output varies from industry to industry, the impact on production imposing a given rate of antipollution tax or user charge will also vary from industry to industry.

Interdisciplinarity has implications for the design and organization of research. Coupling interdisciplinary research with an overriding objective of informing policy-making has direct implications for the research and dissemination process. Successful research, in terms of value for money, calls for a clear framework. This is especially important in an interdisciplinary-research project. However, establishing a clear framework for an interdisciplinary-research project is difficult because various disciplines may indicate different thresholds and patterns of relationship among the variables. For example, in UPLB–UHP, although the team had a good idea of its overall concerns, the research evolved as it was implemented and it was found necessary to bring in people from other relevant disciplines. The mangrove valuation study was probably atypical because the issue was well defined and the potential contributions of various disciplines were more clear cut. The marine scientists looked into the status and role of dissolved nutrients in a primary-growth mangrove area and a degraded mangrove site, they also looked at the biodiversity and marine ecology of the mangrove study sites; the forestry experts examined forestry products; the fishery economist focused on the cost of production and operation of fishponds; the agricultural economist studied the behaviour of fishers' families, etc. Finally, the framework for assessing the social benefits and costs of converting and developing mangroves tied these issues together.

The Philippine ENRAP stands somewhere between the UPLB–UHP and the mangrove valuation project. It had a clear framework in terms of the income-accounting modifications drawn up by the principal foreign consultant, who had done a similar study in the United States. But ENRAP posed challenges because it had far fewer primary or survey data to work with, which affected the project's estimation techniques. The disciplinary experts helped in some of the estimation processes and, perhaps more important, in the preliminary analytical studies that were designed to provide an indication of the potential policy usefulness of ENRAP estimates.

Sajise (1984, p. 320) highlighted key facets of the process involved in interdisciplinary research that "promote the group process of conceptualization of the research design, data collection, synthesis, and evaluation [and] provide the basis for assigning priorities for allocation of project funds." These key facets are the following:

+ Formal and informal group discussions and seminars conducted to smooth and strengthen human and discipline interactions among the researchers;

+ Organization into working groups as needed;

+ Synchronization of activities of the various team members;

+ Adoption of a system of data collection, sharing, storage, retrieval, and synthesis;

+ Periodic internal assessments; and

+ Development of external linkages to obtain views on the research methodology and the policy implications of the research program.

It is apparent from the above that a successful interdisciplinary research project calls for a very good and credible overall coordinator who can facilitate group interaction on the research framework, analysis, and synthesis and properly synchronize the activities of the team. Sajise's 1984 paper was based on his experience as the lead coordinator of UPLB–UHP.

ENRAP also involved extensive coordination, informal discussions, seminars and workshops, a common system of data storage and retrieval, and inputs from outsiders on the research methodology and policy implications. The mangrove valuation project was smaller than the other two research projects, and the organization of the research team was less involved but more specific. However, substantial informal and formal interaction took place among the members of the research team and between the team and selected interested researchers and policy analysts.

The three research projects also involved, in varying degrees, the beneficiaries of the research. ENRAP has had a steering committee composed of key policymakers. But perhaps more important, it has involved technical staff members from the agencies that can provide data or make use of the data from ENRAP. The last phase of the project was the institutionalization phase, in which ENRAP training was given to the concerned agencies. At the policy level, the ENRAP results have been presented to the relevant committees of the Philippine Council for Sustainable Development and in various other forums (for example, congressional hearings, international conferences).

Similarly (but less elaborately), the mangrove valuation project had a steering committee for policy guidance and support; in this case, the committee included the director of the Bureau of Fishery and Aquatic Resources (BFAR) and concerned private-sector representatives. However, funding problems adversely affected the project's work program. As a result, the interaction of the research team with BFAR and concerned private groups was far less than in the case of ENRAP.

In the research dissemination stage, the interaction of the research team with the policymakers and the concerned private groups is expected to be more intense. Both ENRAP and the mangrove valuation study were to some extent undertaken in response to requests from policymakers. Not surprisingly, both projects brought the policymakers into the research process, albeit primarily for broad policy guidance. In the case of ENRAP, the involvement of technical staff, as the source and users of ENRAP estimates, was critical to the success of the research. ENRAP results have also been presented to NGOs.

In the case of UPLB–UHP, policy considerations were somewhat more removed, inasmuch as the overriding objective of the team was to understand the dynamics of upland ecosystems. Nevertheless, it is worth noting that the UPLB–UHP team took the beneficiaries' concerns into consideration. For example, the team agreed (Sajise 1984, p. 321) that the

> position of the program is to relay the results of the study to the community where it was conducted and not to decide for the community what must be done; that if a researcher should assume a stand or a position on an issue resulting from the study, that this position should be the individual's position and not that of the group or the program.

Of course, the personal interactions of the team with key officials at the Ministry of Natural Resources enabled the team to have a key policy impact on the land-tenure issue in the uplands.

Conclusions

From the three examples discussed above, the following appear to be the most important facilitative factors for successful integrated policy research:

✦ *A clear research framework, good research design, and an effective implementation process* — Interdisciplinary research can add substantial value to purely disciplinary research. However, good interdisciplinary research and good disciplinary research can also be complementary. The results of interdisciplinary research may reveal the need for in-depth disciplinary research (as in the case of land tenure in the uplands) or, vice versa, the results of disciplinary research may indicate the need for good interdisciplinary research (as the revision and refinement of ENRAP estimates will entail).

✦ *Good interaction with decision-makers, either through formal mechanisms like a steering committee or through informal discussions* — In the three examples discussed in this chapter, the project leaders were well-known researchers, with good interpersonal linkages with DENR staff. This type of linkage is important for policy-oriented projects. One of the project team leaders noted that several informal, one-on-one discussions are more effective than formal presentations. Also important are informal discussions with various interest groups.

✦ *Good interpersonal relations among team members and an openness to alternative disciplinary perspectives* — As Sajise (1984) pointed out, a strongly individualistic scientist will have difficulty working with an interdisciplinary group. In any case, it takes a conscious effort to ensure continuous, smooth interpersonal relations. Sajise came up with 14 don'ts that will ensure that interpersonal relations of the interdisciplinary team are smooth. In the cases of ENRAP and the mangrove valuation study, the team members had been working with each other on several projects, and they obviously respected one another.

There appear to be two major constraints to effective interdisciplinary integrated research:

✦ *Funding constraints* — Because activities for effective interdisciplinary research must be synchronized, funding problems can disrupt the research timetable and undermine the effectiveness of the program.

✦ *Limited pool of good researchers* — Rare are those who can work together and synthesize effectively the inputs from various disciplines. As such, the risk of poorly executing interdisciplinary research can be high. It must be noted that in the three examples from the Philippines, two of the projects were headed by the leading natural-resource economist in the country, and she is increasingly overworked. However, the other leader is now one of the key proponents of interdisciplinary research on ecosystems, under the aegis of the Southeast Asian Universities Network. Given that the pool of good researchers is limited, there may be a need to take a look at the trade-off between the insights from interdisciplinary research and the risk of undertaking expensive, poorly executed research. Ultimately, only well-done research is credible.

Two major factors appear to affect the policy and program impacts of research for integrated economic, social, and environmental policies.

✦ *The political economy of decision-making* — Major policy decisions go through many processes and involve many officials, both executive and legislative. In many cases, public hearings are required. Hence, research results are probably only one consideration, albeit an important one, in policy decisions.

✦ *Institutional capability to undertake integrated programs* — The experience with integrated rural development (IRD), in vogue in the international donor community in the 1970s, showed that the bias of IRD programs toward a package program (although to be undertaken in phases) overlooked the problem of very limited human and material resources in developing countries (Castillo 1983). As a result, IRD programs failed to deliver as expected and lost much of their initial lustre.

Chapter 10

Concertación: Integrated Planning and Development in Peru

L. Soberon A.

✧ ✦ ✧

Introduction

The transition to sustainable development requires societies to balance social, economic, and environmental objectives. In terms of politics and policy development, this transition can be a negotiation and learning process, with diverse societal groups from government, business, and civil society attempting to

+ Reconcile their various economic, environmental, and social needs and interests;

+ Reach agreement on a common development vision; and

+ Undertake concerted action to realize that vision through both collective action and coordinated individual initiative.

To achieve sustainable development and the integration of social, economic, and environmental policies, we must think in terms of a societal effort at reconciliation and collaboration, with diverse actors accepting and implementing a development program that meets the needs of all. In Latin America, such an approach to sustainable development — participatory dialogue, agreement on a common vision, and collaborative initiative — is known as *concertación*. *Concertación* comes from *concertar*, which means "concert, reconcile, harmonize, bring together, or come to agreement." Like *sustainable development*, *concertación* implies a reversal of the status quo. The call to *concertación* in Latin America has developed in the context of

+ Centralized decision-making;

+ Authoritarian and repressive regimes;

+ High levels of income inequality;

+ Lack of coordination among government bureaucracies, nongovernmental organizations (NGOs), and development organizations;

+ A high degree of social polarization and conflict; and

+ A corrupt political culture in which powerful elites manipulate decisions at the expense of marginal groups with little effective influence.

In the 1980s and 1990s, many countries of the region made progress
toward democratization. Electoral processes took the place of military dictator-
ships. Some countries signed peace accords, marking the end of years of civil
war. Meanwhile, debt crises and structural adjustment dealt a severe blow to
national governments. Many responded through ad hoc decentralization programs, shedding responsibilities to provinces and municipalities without providing the required resources. Democratization, decentralization, and peace
opened up spaces for dialogue and negotiation among diverse and conflicting
social sectors and for formerly excluded groups to discuss their needs and bring
them to bear on policy decisions.

However, tremendous obstacles to *concertación* and sustainable development still remain, such as economic inequality and the hierarchical and exclusionary social and political structures. One of the biggest obstacles to developing
integrated policy through *concertación* is a strongly sectoral mode of policy-
making, which is the result of the restructuring of government into sectoral
departments that do not coordinate their activities. Other barriers include cultural conflicts between communities that follow the traditional values of communal solidarity and economic and social actors that operate on the basis of
modern individualism. These conditions partly explain why concepts like *concertación* and sustainable development so often appear in official rhetoric without being transferred to social practice.

In Peru, progress toward democratization was reversed in 1992, when
President Alberto Fujimori, in an *autogolpe* (self-made coup), suspended the
Constitution. In the early 1990s, the Peruvian military brought under control
the guerilla warfare that had been debilitating Peruvian society throughout the
1980s. Under the new approach to authoritarian rule, some earlier measures of
decentralization were rescinded, but some authority devolved to local government, albeit in a haphazard manner, as the central government shed its responsibilities in implementing structural adjustment.

Nevertheless, Peru is attempting to design an approach to sustainable
development and integrated policy-making on the model of *concertación* (Sagasti
et al. 1995). One notable experience is the interinstitutional *Mesa de Concertación,* supported by the municipal government of the province of Cajamarca.
Mesa (table) is used here in the sense of round table. Conceptually, a *mesa de concertación* is a forum where various groups try to come to agreement on problems
and priorities and coordinate their plans and activities so that, to the extent

possible, they can meet common needs through concerted action. This innovative approach to dialogue among social actors, which is experimented with in various pockets of Latin America, has some resemblance to the round-table approach to consensus decision-making practiced in Canada.

The *Mesa de Concertación* in Cajamarca is "a recent experience in the formulation and implementation of policies and programs through a process of participatory regional development planning" (Mujica 1995, p. 1). Taking part in the process are the municipal government of Cajamara, private industry, NGOs, grass-roots organizations, and regional offices of the different government agencies. This experience, involving about 100 institutions, is "a promising innovation for the learning of inter-institutional collaboration and the formulation of integrated policies," as well as "a good example of an alternative developed at a local government level as an answer to the centralization process currently developing in Peru" (Mujica 1995, p. 1).

This chapter describes Cajamarca's *Mesa de Concertación* and analyzes its potential as a collaborative mechanism for formulating and implementing integrated policies in Peru.

The Peruvian political context

Municipal government in Peru

Peru's political and administrative structure has three levels: departments, provinces, and districts. Currently, it has 24 departments, 188 provinces, and about 1 800 districts. Each province has a provincial council, headed by a provincial mayor. The council is elected by universal, secret, and democratic process for renewable 3-year terms. According to the current *Municipalities Act*, the political party with the most votes gets one-half plus one of the seats in the municipal council; the rest of the seats are proportionately distributed among the other registered parties. This allows the winning party to count on the required majority to effectively implement its electoral platform and ensures sufficient opposition for debate and alternative proposals. The same structure and mode of operation are replicated in the districts.

Centralization and decentralization

Peru's politics and administrative organization have been traditionally centralized. Despite the balance of power (Executive, Legislative, Judicial, and Electoral branches), the Executive Branch traditionally concentrates resource management and appointments (including those in the other branches) and, for all practical purposes, takes the initiative in legislation. With a few exceptions, the ruling party has enjoyed a majority in a Congress ready to bow before the Executive Branch.

Peruvians have protested such centralization and demanded decentralization. The 1979 Constitution responded to these demands in three ways: by democratizing municipal (provincial and district) governments; by establishing regional governments; and by deconcentrating executive administrative powers. Only since the approval of the 1979 Constitution have provincial and district mayors been democratically elected, rather than autocratically appointed by the Executive Branch. The 1984 *Municipalities Act* grants provincial and district governments a wide range of responsibilities for governance, although municipalities receive a very limited amount of resources to manage such responsibilities.

The Acción Popular Revolucionaria Americana (APRA, American popular revolutionary alliance) administration of President Alan Garcia (1985–90) gave decentralization strong support, as mandated by the 1979 Constitution. This government organized a regional level of administration, intermediate between the departmental and national levels. In total, it organized 13 administrative regions, each with its own regional assembly. Representation in the regional assemblies was divided equally among the provincial mayors, representatives of political parties, and representatives of grass-roots organizations. The presidents of regional assemblies were elected by popular vote in each of the regions. Through the regional assemblies, "local interests were given representation at the top political decision-making echelons in the region" and "responsibilities were shared by politicians and grass roots leaders." This arrangement offered improved scope for "the integrated design and implementation of policies" (Mujica 1995, p. 5).

During the 1980s some progress was also made in devolving public administration. The ministries, all headquartered in Lima, assumed normative functions. Executive functions were given to deconcentrated agencies and provincial and district operational units. Regional secretariats coordinated these agencies' functions.

Despite these reforms, ministry headquarters retained wide control of governmental affairs, a situation accentuated during President Fujimori's administration. During the *autogolpe* of Fujimori's first administration (1990–1995), the government suspended the regionalization process. Executive Branch appointees temporarily replaced elected officials. The Fujimori administration also initiated "a campaign against municipal governments which both discredited mayors and cut their financial resources," a political move to centralize "decision making and fund management to an even greater degree in the Executive Branch, in particular through the Ministry of the Presidency" (Mujica 1995, p. 5). In addition to their impact on the growth of democracy and the likelihood of integrated programs being implemented at the regional level, such policies also resulted in the confrontation between the state ministries and the municipal governments. Municipal functions were gradually cut or assigned to the ministerial agencies that funneled financing.

A characteristic of present-day politics in Peru is a strong tendency to concentrate decision-making power in the capital city of Lima, especially in the Office of the Presidency. The country may be said to live under the sign of centralized decision-making and resource distribution, although provincial and district administrations may support decentralization.

Policies

Under the current administration's structural-adjustment program, market forces are expected to allocate resources; economic agents, to gain greater efficiency and competitiveness in an open economy; and the state, to create a suitable legal framework and provide the necessary guarantees and stability. Foreign investment plays a key role in this vision of economic development. However, substantial efforts are made to promote large-scale domestic savings.

A streamlined government now features fewer tiers of coordinating and planning agencies and correspondingly stronger sectoral policy-making. The notion of a ministerial cabinet can even be said to have lost vigour. Government actions are seen as stemming from the president. A new institutional structure is emerging around the notions of overseeing economic players and preventing trusts, market dominance, fraud, and corruption, through the use of sectoral controls.

The administration's main achievements have been restoration of peace, reduction of inflation, reintegration into the international financial community, modernized and enhanced tax collection, and positive rates of GDP growth for several years in a row. As a whole, over the last 5 years, government has managed to turn the low morale of most of the population, which had been overwhelmed by a feeling of total collapse and catastrophe, into a positive feeling, although some people still voice qualifications and specific complaints about political and social issues.

Government argues that the advances made will permit it to pay more attention to the social aspects of development, education in particular. President Fujimori's inaugural speech announced his administration's support for family planning, reflecting government's intention to include population considerations from now on in public policy-making.

Peru's major challenge is to generate jobs on a massive scale. After a long period of economic deterioration and intense growth of the working-age population, unemployment has become a huge problem. Poverty must be brought under control, and it is necessary to reduce the extreme poverty that afflicts a large part of the population, especially in the rural highlands. Likewise, public education must be rescued from its current state of decay, and the trade gap has become an increasing cause for concern. Consequently, environmental issues command little high-level attention.

Cajamarca: an overview

Located in the northern Andean highlands, the department of Cajamarca has a predominantly rural population and is one of the poorest departments in Peru. Moreover, Cajamarca has one of the highest relative rural-population rates and a negative migration balance.

The department of Cajamarca is divided into 13 provinces. Its capital is the city of Cajamarca in the province of Cajamarca, itself comprising 12 districts. According to the 1993 national population census, the province of Cajamarca had for that year a total population of 236 500, of which 59% belonged to the rural sector, 5% lived in minor urban centres, and 39% lived in the city of Cajamarca (92 447 inhabitants). For the countryside, the 1994 national agricultural census gives the following picture: economic units with less than 3 ha represented 60% of the total units and accounted for 7% of the total agricultural area; economic units in the range of 3–10 ha represented 38% of the total units and accounted for 40% of the total area; and economic units larger than 10 ha represented 2% of the total units and accounted for 53% of the total area. This land-tenure structure is associated with *minifundio*, on the one side, and with small and medium-scale commercial farms, on the other.

The economy of the province is to a large extent based on milk production, organized around the Nestlé company's evaporated-milk operations. The company arrived in the area some 50 years ago and made dairy production dominant in the Cajamarca Valley. Current changes in the country's economy (suspension of subsidies; opening up of the economy) have led the industry to a crisis, with adverse impacts on the whole population. Potato production is another important component of the local economy and culture; however, over the last 15 years, potato areas have diminished and yields have stagnated in relative terms.

Cajamarca's mines have traditionally been seen as small, low-technology operations that lost their significance after the recession that struck the mining industry. In recent years, foreign interests and Peruvian mayors have started a new cycle in mining. State-of-the-art gold-mining technology has yielded high returns on investments in just a few years of operation. Currently, the main concern in Cajamarca is to retain part of the income generated by the mining companies and use it to fund local development projects. The most important mining enterprise is Compañía Minera Yanacocha, which is dedicated to the exploitation of gold deposits. Because of foreign mining technologies and hiring procedures, people believe that mining will have no considerable direct impact on the local production structure. Therefore, expectations focus on collecting a mining concession tax (Frias 1993; Amat y León 1994; Villarán 1994).

The Frente Independiente Renovador

In 1993, the election victory of the Frente Independiente Renovador (FIR, independent renovation front), led by Luis Guerrero, marked a dramatic political shift in the province of Cajamarca. The inauguration of the FIR government brought an end to the long APRA regime in the provincial municipal government. The FIR administration faced issues of stagnating production, scarce investment, unemployment, environmental degradation, and some of the lowest living standards in Peru (Mujica 1995). In its new general-policy outline, the FIR administration included the following components:

+ *Modernization of the municipal institutions*

+ *Operations within the framework of sustainable development including environmental considerations as core components*

+ *Constructive dialogue with all private and public institutions at the* Mesa de Concertación *to reach agreements on both rural and urban development projects*

+ *Democratization and decentralization of power*

+ *Incentives to private investment*

+ *Enhanced external relations, both in Peru and abroad, and promotion of tourism*

JVM (1994, p. 7)

With this policy outline, Guerrero's administration has striven to build a new image for Cajamarca. Guerrero pronounced Cajamarca the first environmentally friendly municipality of Peru, launched a strategy of *concertación*, and started an active campaign to build foreign relations and contacts.

An important component of the new administration is its strong links with a network of NGOs, professionals, and academics who are aware of the advantages of a sustainable-development approach that emphasizes environmental awareness, public participation, collaboration with district governments, and an ecosystems approach to integrated watershed management. This network was based on contacts that Guerrero, an agricultural engineer, had made while conducting university research and working with NGOs and grass-roots organizations. This network also had the professional and political capacity to obtain financial and technical resources through contacts with international technical-cooperation agencies.

Although the lessons learned so far have stemmed from local development in the relatively small areas of influence of NGO projects, they may reach a broader scope of influence, thanks to the support and collaboration of this new government. FIR is expanding the role of the municipality, from one of delivering urban services to one of promoting regional development. This marks a substantial departure in municipal governance, not just in Cajamarca but throughout the country.

Decentralizing municipal government

Despite the fact that the Province is subdivided into 12 district municipalities, each with its own elected mayor, it was still difficult for the voice of people at the grass-roots level to be heard by the municipal government of Cajamarca (Joseph 1997). Guerrero therefore made use of a rarely exercised provision of Peruvian municipal law to found a new level of representative government, below that of the districts. According to Mujica (1995), Peruvian municipalities rarely use this mechanism, as the provincial councils and mayors wish to retain direct decision-making power. Guerrero's administration created 12 new neighbourhood councils within the city of Cajamarca, as well as 42 new rural councils in rural areas of the province. In July 1993, each council's representatives were elected by direct universal vote (Joseph 1997). Certain local-government functions were to be devolved to these councils. These new authorities, which are mostly grass-roots organizations, now share decision-making responsibilities with the provincial municipality and the district municipalities (Mujica 1995; Joseph 1997).

Municipal-government organization

The municipal government of Cajamarca is divided into the provincial municipality, the provincial municipal council, the municipal board, and the deconcentrated municipal agencies. The provincial municipality comprises the district mayors assembly, the neighbourhood assemblies, and the small townships assemblies, as well as the interinstitutional *Mesa de Concertación*. The provincial municipal council comprises the following 10 municipal commissions, each chaired by a councillor:

+ Transportation and traffic;

+ Education, culture, and sports;

+ Development promotion;

+ Community services;

+ Citizen participation and local government;

+ Institution-building and interinstitutional *concertación*;

+ Planning and budget;

+ Women and family affairs;

+ Environment; and

+ Tourism, industry, and crafts.

The municipal board comprises other boards charged with implementing projects on themes such as environment and population, development promotion, and services and culture.

The *Mesa de Concertación*

A central element in FIR's program is a strategy of *concertación*, involving dialogue among representatives of a broad range of public and private institutions and community groups. Debate is used to help identify and negotiate priorities for investing scarce resources in urban and rural development according to a common agenda (Mujica 1995).

On 12 April 1993, the province made its first concrete achievements in *concertación*. The provincial municipality and the various institutions signed 60 frame agreements for interinstitutional cooperation, with the purpose of working together to solve specific problems related to the province's development (Mujica 1995). These agreements form the basis for the participation of a broad range of organizations in the *Mesa de Concertación* and ensure a strong relationship between the *mesa* and the provincial municipality.

Goals and mandate

The vision of *concertación* in the Cajamarca development plan involves "bringing together the will of the people, and their economic, political, social and cultural interests; it requires leaving aside individual attitudes in favour of shared goals; *concertación* is a meeting point of civil society and the State" (MDC 1994, pp. iv–v). *Concertación* also requires "a commitment to make the agreements come true," as "only thus will planning cease to be centralized" (MDC 1994, p. vi). The action and planning strategy of the provincial municipality is to bring together and coordinate "the efforts of public and private organizations with operations in Cajamarca that have signed agreements with the Municipal Government" and whose "actions materialize at the *Mesa de Concertación*" (JVM 1994, p. 43).

The mandate of the *Mesa de Concertación* includes (JVM 1994, p. 43) "drafting diagnoses, policies, plans and projects for the development of Cajamarca," as well as a "provincial development plan for the short and medium terms," to achieve the following specific objectives:

> *(a) To increase the efficacy of development programs through the definition of effective and strategic impact priorities;*
>
> *(b) To increase efficiency through coordination among the institutions working in the province, thus optimizing human and economic resources and avoiding duplications;*
>
> *(c) To build up participation channels for citizens, delegating to them decision making, management and control capacities;*
>
> *(d) To support the decentralization process from the provincial government towards districts, neighbourhoods, and small towns;*
>
> *(e) To reactivate economic growth but taking into account the protection of natural resources, which implies a change of the traditional economic concepts; and*
>
> *(f) To integrate the environmental dimension into the decision making of proposals and programmes for development.*
>
> Mujica (1995, pp. 7–8)

Thus, the provincial government has entrusted the *Mesa de Concertación* with drafting a sustainable-development plan for the province.

Through *concertación*, the various institutions and community groups are expected to propose development projects and initiatives that will respond to their social, economic, and environmental needs. However, in drafting and implementing a plan, the municipal administration and other participating institutions have different roles. The municipal government undertakes promotion and coordination; the participating institutions take the initiative in identifying and implementing priority tasks (MDC 1994). Public and private institutions are expected to implement the development plan in their respective areas of competence, using their own resources. Thus, the institutions are expected to arrive at a consensus on their specific areas of interest and together seek to carry out the projects at the top of their priority lists. This procedure is designed to prevent unnecessary overlap, use individual resources fully, and tackle problems in the framework of a comprehensive development plan.

The *Mesa de Concertación* is thus the current municipal administration's central policy point. It implies an institutional renewal to achieve a more democratic provincial government. The province has great expectations of its efficacy.

Organization and operations

The *Mesa de Concertación* actually consists of a number of *mesas*. Six *mesas* are oriented to specific thematic areas, and a general *mesa* serves as a plenary round table. The thematic *mesas* focus on the following areas:

+ Urban environment;

+ Natural resources and agriculture;

+ Production and employment (manufacturing and handicrafts);

+ Cultural and tourist heritage (rural and urban);

+ Education (rural and urban); and

+ Food, health, and population (rural and urban).

Municipal commissions give technical support for certain thematic *mesas*, including those working on the urban environment and food, health, and population. Various other organizations participate in the thematic *mesas* of interest to them. Each thematic *mesa* holds monthly meetings (chaired, until recently, by a councillor who also chairs one of the municipal commissions). Some councillors participate in more than one thematic *mesa*. Most of them participate in the urban-environment *mesa*, but their participation is more evenly distributed among the other *mesas*.

Although the provincial council previously appointed councillors to the chairs of the *mesas*, the thematic *mesas* began to elect their own chairs in May 1997. This measure was meant to give greater autonomy to the *Mesa de Concertación*, to encourage more active participation, and to ensure that the *mesas* had the best-qualified individuals as chairs. In only one case was a councillor who had previously occupied a chair elected to the same position; in the other cases, the *mesas* elected representatives of organizations working in their respective thematic areas.

Coordination among the thematic *mesas* is achieved in several ways. First, the presence of councillors ensures that the provincial council is kept aware of the discussions and activities in each thematic area. Second, the councillor who chairs the municipal commission for institutional-building and interinstitutional *concertación* has specific responsibility for the general coordination of the thematic *mesas* (Mujica 1995). The councillor is supported in this task by a facilitating team comprising a variety of local and Lima-based experts; the *Mesa de Concertación* Office of the provincial municipality also supports these coordination activities (Mujica 1995). Finally, the members of the thematic *mesas* meet in the general *mesa* once every 6 months.

The general *mesa* is chaired by the provincial mayor. Coordination is generally the duty of the councillor who is president of the commission for institutional building and interinstitutional *concertación*. At the meetings of the general *mesa*, members define the general objectives, main lines of activity, and the overall annual working program and plan concrete actions to further the drafting and implementation of the provincial development plan (Mujica 1995).

The *Mesa de Concertación* is an informal association, with no legal or institutional status or statute governing its operation, apart from the frame agreements between the participating institutions and the provincial municipality. Although meetings have a minimum of formality, certain conventions are followed: meeting agendas are announced in advance; debates are directed by the chair; and decision-making is based on consensus and follows the consensus-based practices used by the NGOs in their planning exercises (international donors and government agencies also favour consensus decision-making). Consensus decision-making often involves prolonged processes of debate and discussion of alternatives. The chair or facilitator of the *mesa* takes note of points of agreement and summarizes them. If there are no objections, the *mesa* makes a decision.

Planning process and methodology

The planning process started within the thematic *mesas*. The *mesas* undertook diagnostic and planning workshops, using the planning-by-objectives method of the German Technical Corporation. This method follows a participatory approach and is meant to generate a strong commitment to a plan from institutions, elected officials, and authorities, as the objectives and implementation strategy for the plan are to be based on a common understanding of the issues and possible solutions.

The thematic *mesas* drafted working documents with the following components: problem diagnosis, objectives, intervention strategies, and project identification. The documents were prepared with varying degrees of sophistication, depending on the experience of the thematic *mesa*. To contribute to the mid-term plan outline, each thematic *mesa* presented its working document at a seminar, "Cajamarca: Democracy, Environment, and Development" (Mesa de Concertación 1995).

Problems in the planning process included little experience in using the methodology; inadequate information and technical support; and limited cross-referencing and thematic integration of diagnoses and proposals. These problems reveal that *concertación* and planning involve a challenging learning process and that policy integration is thus likely to be achieved slowly and incrementally. Also, until the seminar, the entire plan-drafting process, from problem diagnosis to project identification, had taken place in the thematic *mesas*; it is unclear what avenues were left open to enable the thematic *mesas* to develop a common understanding and reconcile their priorities. Iterations of working-group and plenary sessions might have been beneficial to the process.

Although the *Mesa de Concertación* has undoubtedly made considerable progress in defining policy outlines for the sustainable-development plan for Cajamarca, much of the work of integrating policies appears to be lacking. The challenge before the *mesa* is to structure the proposals of the thematic *mesas* in a comprehensive and integrated development plan.

Short-term project *concertación*

For project implementation, the *Mesa de Concertación* proposed that each institution continue to perform its respective tasks in the field, using its own resources but coordinating its activities with those of the other institutions. Initially, the leaders of many participating institutions misunderstood this mechanism and perceived the *Mesa de Concertación* as a source of funding. Once the *mesas* got over the disappointment of finding this was untrue, they were able to take part in defining policy guidelines for the sustainable-development plan and also devise a number of projects to reflect their own interests.

Response to *concertación*: involvement of the various sectors

In general, the *Mesa de Concertación* in Calamarca has succeeded in bringing together both private and public institutions. About 100 institutions are involved in thematic *mesas* — a broad-based and sustained institutional mobilization. However, it is apparent that some sectors' involvement in the *Mesa de Concertación* has been restricted by several factors.

Because government agencies are regulated, they may be reluctant to take on additional responsibilities. The *Mesa de Concertación* found that the agency heads refuse to attend meetings; instead, they send their representatives, who lack the authority to commit their institutions. For instance, the thematic *mesa* working on natural resources and agriculture has only the limited participation and interest of the Ministry of Agriculture and related governmental agencies. They do not regard the *Mesa de Concertación* as a promotional tool and rely instead on direct relationships with the involved players.

The Universidad Nacional de Cajamarca (national university of Cajamarca) has responded in various ways. Some departments have chosen not to become directly involved but take a watchful stance. Other departments are actively involved, mainly those with administrators and professors who are councillors and chairs of thematic *mesas*.

Labour has shown no positive response. Some unions refuse to take part in interinstitutional discussions or commit themselves to agreements, preferring to be fully independent of the process.

The business community has shown a disjointed response. The local chamber of commerce is a weak institution. The thematic *mesa* concerned with production and employment offered to help build this institution, but the institution finds it difficult to commit itself to the decisions adopted by the *Mesa de Concertación*. On the other hand, the hotel and restaurant industry has reacted more vigorously and actively, showing a capacity to discuss its needs, make proposals, and reach some agreements.

In the banking industry, local bank managers lack the authority to participate in the *Mesa de Concertación* and depend heavily on their Lima headquarters, which have little interest in regional development. The *Mesa de Concertación* made an effort to draw the interest of the Caja Rural, an organization that operates as a rural savings bank, to channel loans to producers' committees. Through the rural savings bank, a direct relationship was established between the municipality and the Livestock and Dairy Fund. However, the participation of the livestock and dairy sector in the *Mesa de Concertación* remains limited. Some dairy farmers maintain that their main problem is the low price of milk and that the solution will be found only through direct talks with the central government and lobbying in Congress, and they fail to see how the *Mesa de Concertación* can promote their goals.

In contrast, the Servicio Nacional de Adestramiento en Trabajo Industrial (SENATI, national technical training service) has reacted very positively and is developing various joint agreements. This is due, among other reasons, to the fact that the councillor who chairs the thematic *mesa* on production and employment supported the creation of the Cajamarca SENATI institute.

Among all the mining companies operating in Cajamarca, only the Yanacocha Mining Company participates in the *Mesa de Concertación*. Yanacocha participates in the thematic *mesa* concerned with production and employment and collaborates in projects on light, water, and sewage systems expansion.

Of all sectors, the NGOs responded perhaps the most enthusiastically to the municipality's call, especially those in any way connected with FIR. On the other hand, NGOs unconnected with municipal politics have had a less clearly defined attitude. Despite their participation in the *Mesa de Concertación*, their commitment remains low, in part because they perceive discriminatory treatment against them and in part because of political differences. Furthermore, they lack access to information about the activities of the *Mesa de Concertación*. Still, they support the concept of *concertación*.

Cajamarca's *Mesa de Concertación* has generated interest and expectations within the technical-cooperation community. A large contribution to this was the municipality's dissemination effort in Peru and abroad, together with the *mesa's* record of concrete achievements in the region. It is expected that the *Mesa de Concertación* and the sustainable-development plan will become the framework for technical cooperation and facilitate interinstitutional agreements, as well as follow-up and evaluation of development actions.

The political opposition has made a positive assessment of the concept of *concertación* but disagrees with the municipality about the way consensus should be built and the way projects should be implemented. It wants more participation of grass-roots organizations in the *Mesa de Concertación* and wants the *mesa* to operate more openly, with less influence from the NGOs connected with the mayor. Opposition parties see the development plan as merely the sum of specific proposals designed to respond to the particular interests of each NGO. Opposition parties also charge that the municipality established the district *mesas* on the basis of political rather than technical criteria, with a view to winning the coming election.

Potential and challenges

Potential

From the standpoint of integrating economic, social, and environmental policies, Cajamarca has some potential. This is a region where rural-development NGOs first adopted an agroecological perspective and where many projects with an ecodevelopment orientation have been carried out. Environmental concerns are prominent on the development agenda of local municipal administrations and the leading NGOs participating in the *Mesa de Concertación*. Moreover, the environment is a central priority among the local academic and technical communities, and as such it has been included in the development efforts under way in the microwatersheds of the province.

The *Mesa de Concertación* has managed to draw the interest of many public and private institutions and spurred the largest ever institutional mobilization, to draft the sustainable-development plan. This process involved representatives of diverse institutions, interests, and professional fields. The *Mesa de Concertación* provides the space for the intersectoral and interdisciplinary debate and discussion that are fundamental to conceptualizing and formulating integrated policies; it also provides an invitation to take part in the collaborative process that is required to implement them.

Challenges

The *Mesa de Concertación* is the type of interinstitutional mechanism that seems to be needed to achieve integrated policies and planning. But several factors related to the structure, operations, and participation have inhibited this achievement. From a basic institutional standpoint, a more fundamental problem with the *Mesa de Concertación* is its lack of autonomy, or its dependence on the municipal government. From the observations made, the key strategic challenges for the *Mesa de Concertación* are to achieve greater institutional autonomy and to improve opportunities for learning the process of integrated policy and planning.

Institutional autonomy

Although at present all players conceive of the *Mesa de Concertación* as an interinstitutional collaborative mechanism, the municipality weighs in heavily. The mayor chairs the general *mesa*, and councillors have, until recently, chaired all the thematic *mesas* and also headed the municipal commissions. The resulting strong coordination with the municipal government seems a positive feature, but the excessive influence of municipal authorities may hamper efforts by the *Mesa de Concertación* to bring together the multiple and diverse social, economic, and political actors.

At the outset, the councillors who headed the thematic *mesas* belonged to FIR, except for one. A situation emerged in which the mayor and his political entourage exerted control over a thematic *mesa*, which led other political parties to keep their distance or to take a less active role. Tight control over the higher positions in the *Mesa de Concertación* has excluded the leadership of other individuals and institutions, along with their experience and initiative. It will be interesting to observe the impact of having elected chairs in the thematic *mesas*.

Another component of this complex situation stems from the mayor's and FIR's connections with the NGO sector, which gave birth to the concept of and impetus for the *Mesa de Concertación*. NGOs affiliated with the mayor's entourage were also instrumental in starting the *Mesa de Concertación*, making it initially operational, and even establishing a technical-support office. Consequently, they took a high profile in organizing and running the *Mesa de Concertación*. Interviews revealed a certain sensitivity about this.

The implication is that the institutional nature of the *Mesa de Concertación* needs to be more clearly determined. One option is to make it an institutional mechanism within the municipal administration. Alternatively, it could be turned into an autonomous institution, putting the municipal government on an equal footing with the other public and private institutions.

Policy integration

The experience with Cajamarca's *Mesa de Concertación* sheds light on several issues of importance in the integration of social, economic, and environmental policies. As a collaborative mechanism that brings together diverse sectors of society, it is an example of an institutional mechanism with the potential to formulate and implement integrated policies.

How does the approach of the *Mesa de Concertación* to decentralization and local participation aid the process of policy integration? To the extent that it includes groups that can articulate social, economic, or environmental concerns and it is able to integrate those concerns into plans and actions, the *Mesa de Concertación* has the potential to effectively integrate policy and promote sustainable development. However, little progress has been made in developing an integrated-policy approach. Most approaches still focus on specific sectors, despite requirements to coordinate sectors. The principal directors of the *Mesa de Concertación* recognize that relatively little progress has been made toward an integrated approach to planning and policies.

This has been under discussion, however, as people have some awareness of the need for integration. For example, education programs are needed to reinforce the regeneration of forest cover, and this should be linked to social and economic benefits (forest cover in the upper reaches of the watershed is important to maintaining water flow and supply to urban areas). Most important, it should be recognized that integration involves an incremental learning process: recent experience, such as that of the Cajamarca *Mesa de Concertación*, shows that integration can only proceed gradually.

To what extent can the vertical integration of the various levels of government be achieved? A key obstacle to the development of integrated policies through the *Mesa de Concertación* has been the delegation of junior representatives from government agencies who lack the authority to make decisions for their organizations. This has led to a situation in which NGOs take the primary initiative in decisions, and the onus is on the peripheral government representatives only when there is a need to further an initiative in a given policy direction. This and subsequent observations of the Cajamarca case have also raised the normative issues of how integration is to be managed and by whom.

The organization of the *Mesa de Concertación* into thematic *mesas* has resulted in a sectoralized mode of operation, with a marked division of labour. This is partly because the various mechanisms to coordinate the thematic *mesas* (mainly forums and workshops) have been inadequate to the task or ineffectively used. Indeed, interviews indicated that global workshops for general debate and formulation of policy priorities tend to be organized in working groups that reproduce and reinforce divisions along thematic lines. As a result, the priorities of diverse sectors appear in formulations of specific policies and in the provincial sustainable-development plan, without having entered into a debate on overall priorities, which would involve the reconciliation of interests. This point came up at a recent workshop in which a global prioritization was attempted. However, workshop participants also found that certain priorities were emphasized across thematic areas, such as education and employment, which suggests the need to ensure that participants comprehensively integrate these objectives into their work on the various themes.

Much of the task of integrating economic, social, and environmental policies has been restricted by the scope of the thematic *mesas*, although there is discussion of how priorities should be set within the projects of the member institutions that implement the sustainable-development plan. Thus far, the greatest impact of the debates, studies, and articulation of priorities in the *Mesa de Concertación* has been on the institutions' plans and policies, rather than on their activities.

Furthering an integrated approach to development through *concertación* in Cajamarca requires considerable effort from the *Mesa de Concertación* to promote and support the formulation of integrated policies by enhancing opportunities to learn integrated modes of developing and implementing policies and programs.

PART IV

CONCLUDING PERSPECTIVES

Chapter 11

Integrating Environmental, Social, and Economic Policies

S. Holtz

❖ ✦ ❖

Introduction

As the essays and case studies in this book demonstrate, one can look at the attempt to integrate environmental, social, and economic policies or goals from a number of distinct perspectives. Conceptually, these various viewpoints fall into three main ways of framing the issue:

+ Seeing integration as an intermediate, process-related goal, with the ultimate end being a better balance among these three policy areas (for example, DePape's and Penfold's studies);

+ Seeing integration primarily as a process goal (for example, Bernard and Armstrong's study); and

+ Seeing integration as a substantive goal, the major challenge being to identify and resolve contradictions and difficulties stemming from the interrelation of subsidiary goals in these three policy fields (especially Robinson and Tinker's study).

Individual authors can, of course, see integration from more than one of these perspectives. This categorization merely helps clarify and put in focus some of the major themes in this discussion, rather than being an attempt to pigeon-hole articles on this complex subject. Each of these ways of seeing integration offers some insights and has some blind spots; each has, either implicitly or explicitly, its own measures of the success or failure of integration; and each has some unresolved issues, whether in terms of other frameworks or in terms of its own. This chapter considers a number of issues in integration from an historical perspective and from each of these three perspectives and draws some conclusions about the most useful directions for future discussion, research, and analysis.

Integration as a part of the history of the debate on sustainable development

Integrating environmental, social, and economic policy is discussed in connection with sustainable development, a concept that goes back to the 1970s and earlier but came into prominence in the public-policy arena in the late 1980s with the release of the Brundtland Commission report (WCED 1988). At that time, the related policy questions were primarily about the perceived and real tensions between environmental and economic objectives. Social considerations related to equity, political inclusion, and discrimination were just beginning to enter this discourse.

During the late 1980s and continuing into the early 1990s, most people supporting sustainable development saw the first objective in integration as being simply to bring the three areas of policy and decision-making into closer contact to enable them to inform each other, rather than allowing each to continue to be a separate policy stream. Many commentators have pointed out that compartmentalizing environmental, social, and economic goals can and does lead to policies in one area undercutting those in others, such as (to take a Canadian example) certain agricultural-subsidy programs that had the unintentional effect of accelerating the destruction of wetlands; environmental programs to address this loss of waterfowl habitat also cost significant amounts of public money (as discussed in Sopuck [1993], for example).

More fundamentally, however, the problem of policy integration appeared to be one not merely of avoiding such obviously perverse effects but also of achieving a new balance of policy goals. This meant not only bringing these separate areas of policy into closer contact as policies were developed but also increasing the emphasis on environmental and perhaps social objectives in policy settings, which usually seemed to be dominated by economics. However, even then, people differed in the emphasis they put on the interpretation of this common objective.

At that time, some environmentalists were committed primarily to advancing environmental concerns and worked on these issues as volunteers or professionals associated with environmental organizations. But other environmental advocates, some recently converted, worked in business, industry, government, churches, unions, and international development and nongovernmental organizations (NGOs), and they continued to see economic, political, and social considerations as the primary factors in their decision-making frameworks, even as they began to acknowledge the significance of the environment. For this latter group, the huge practical challenge at the heart of sustainable development was to find ways to integrate new factors and objectives into decisions made in the daily course of keeping an organization, business, or government department functioning in its existing role. For many environmental activists, having the environment become an important factor in actual policy decisions in such sectors as forestry or energy was only part of a larger goal of reorienting society's underlying values and thinking. For these environmentalists, education was of the essence; their ultimate goal was to bring about a philosophical shift away from a short-term, mechanistic, and purely human-centred perspective to a long-term, ecological, and biocentric one. This is not to say, of course, that these two groups were absolutely distinct but simply to emphasize that people's ways of seeing policy integration depend on their goals, and these differ even among people marching quite sincerely under the banner of sustainable development.

Nevertheless, from both pragmatic and philosophical points of view, the basic problem at the beginning of the debate on sustainable development was to find ways to insert new values, objectives, and information into existing policy- and decision-making processes to alter their outcomes. New institutional arrangements and tools were needed. People with different agendas would have to gain access to decision-making processes; new kinds of information would have to be generated and actually brought to bear on the issues; and better tools for analyzing and evaluating complex situations and outcomes would be helpful, if not essential. If the same decision-makers continued to do business in the same ways, they would produce the same types of policies and decisions and their existing priorities would remain fundamentally unchallenged.

Policy integration as an intermediate, process-related goal

This was the background, then, to the first category of viewpoints, that is, the view that policy integration is ultimately about substantive changes in policy direction but requires new decision-making processes and decision-aiding tools. This was the earliest perspective on the issue, and, I suspect, it remains dominant, particularly among people with some responsibility for implementing sustainable development and for activists who want specific decisions and policies to better reflect social or environmental values.

Discussion

It is worth noting again that in this perspective, the point of emphasizing processes and tools is to affect the substance of decisions. Despite this fact, it is easy to gain the impression that within this perspective, process can take over as the focus, even though the real disagreements are substantive. Often issues of design and fairness of a process generate intense controversy, even before any decision is made. Criticism of the process can become a tactical option for delay or an easy way to call the legitimacy of any decision into question without having to make a convincing argument based on hard evidence. Conversely, people sometimes point to the elaborateness of a decision-making process as being the main justification for its outcome, rather than the quality of the work.

Despite these tendencies to downplay the substantive results of policy processes designed to better integrate environmental, social, and economic factors, some progress has unquestionably been made. In the decade or so since the Brundtland report, interest has increased tremendously in tools and institutional processes that can be used for policy integration. Whole new processes have been designed to bring together people with strong interests in one or another set of values pertaining to an issue to negotiate a solution that all can at least live with. In Canada alone, the list of such institutions and processes is extensive; just focusing on new institutional models, the list includes dozens of local, provincial, and national round tables and similar bodies, such as the New Directions group, which began as an informal organization of business and environmental leaders in central Canada. Other countries have produced their versions of organizations with explicit mandates to further policy integration or sustainable development. Similarly, one-time, special public processes for consultation and negotiation have become routine for developing new legislation and regulations, as well as for addressing particularly controversial issues, such as conservation designations in land use.

Such efforts at integration rely chiefly on opening up the decision-making process to input from a wider range of people. Through this process of inclusion, new information and perspectives are (it is assumed) brought to bear on the outcome, and a better balance of social, economic, and environmental concerns is achieved. When conflict occurs, its resolution (and the substantive outcome) depends on the resources and skills of the people involved, especially time, money, and research capability. In this model, effective mediation and negotiations are the actual tools for integration; integration itself consists of achieving consensus among people with differing perspectives.

By contrast, other models for integration depend less on the mix of people involved in the process and more on certain new information requirements or analytical tools. These usually involve the prescribed development and application of information that previously was not routinely used in the decision-making process. This type of initiative is most common in making routine policy and management decisions in business, industry, and government agencies, where opening up the process to new participants is neither workable nor particularly desirable. Examples include the development of internal environmental-impact assessments; new requirements for purchasing criteria, briefing documents, and the like; setting out information such as a statement of impacts on certain social groups or the environment; and environmental audits and similar check lists.

Analytical tools are more directive: they not only bring forward information but attempt to prescribe, or at least clearly articulate, exactly how different factors are to be weighed. Some of these tools, such as cost–benefit and risk analyses, are not new but may be used in new situations to help decision-makers more clearly lay out the environmental, social, and economic impacts of alternatives. Other tools, such as scenario analysis and game theory (to better explore alternatives in policy decisions) and social and environmental costing and pricing are relatively recent developments. How often they will actually be used to make more integrated policy decisions is still unclear.

An assessment of strengths and weaknesses

Seen from this first perspective, policy integration involves quite modest expectations: promoting a better balance among environmental, social, and economic goals; and making some actual progress toward an acceptable compromise if these objectives clash. This perspective's great strength is its ability to actually move sustainable development forward in both major and routine decisions of governments and corporations.

It has two weaknesses, however. First, and most important, from a pragmatic point of view, it usually assumes that the fundamental structures of power and authority in a society and the motives and intentions of the most senior decision-makers are outside the bounds of the new policy-integration process. The process of integrating these values is undertaken by other actors, presumably for the benefit of these senior decision-makers, such as chief executive officers and government ministers. This is a perfectly reasonable assumption about the real world: if the process has the support of those senior decision-makers, more integrated policies and decisions may well be adopted. This is true for a number of the case studies in this book. However, no guarantee can be given against a complete repudiation of the new policy direction, either through a change of heart or through a change of leadership.

Several of these case studies point to external factors — essentially political in nature, such as new legislation or intense public controversy — that make a situation ripe for leaders to initiate a new, more integrative approach to policy. Such pressure indicates that at least in some quarters of society, people want to rebalance the existing emphases in policy- and decision-making. But adopting an agenda that integrates social, economic, and environmental values, however effectively and elegantly, may still fail to be of the first importance in policy decisions, simply because it is not the main goal of senior decision-makers. The number-one priority that governments of all kinds have is staying in power. This does not necessarily indicate a cynical indifference on the part of politicians; a government must retain power to carry out any program at all. Thus, a government's perception of the political results of a given policy or decision, especially as affecting its chances of keeping power, will override any new integration of social, economic, and environmental interests worked out by others. In the corporate sector, what is likely to trump other factors in final decisions is the perceived impact on a company's economic performance, as well as legal and other risks to viability. In short, to be implemented, policies and decisions that integrate social, economic, and environmental policies must be consistent with reality as seen by senior decision-makers. And that reality is the product of the structural factors controlling the various institutions in society.

Second, this perspective, aiming as it does at a balance of values rather than at a complete reconciliation of apparently contradictory viewpoints, assumes that a positive outcome is always possible. Within this framework, and especially from a practical point of view, this is true if it simply means being satisfied with progress, rather than striving for a perfect solution. However, often there is little in this approach that can help observers (as opposed to actual participants in the process, who are in a position to assess the strengths of opponents and their arguments) objectively determine whether an outcome is good enough. As well, because at least some progress can usually be made in simultaneously addressing goals in all three policy areas, this approach may fail to adequately identify or explore the really difficult policy problems, some perhaps amounting to outright impossibilities.

Policy integration as a process goal

The second perspective on policy integration is that its fundamental goal is to make policy- and decision-making processes more integrative in addressing social, economic, and environmental values and objectives. Historically, this perspective frequently developed from working with the approach that promoted changes to decision-making processes in order to make substantive changes in policy direction. An example of this evolution is the development of environmental mediation, a new field in the 1980s; in many cases, the early practitioners had a strong environmental perspective and were interested in creating ways to get movement on a particularly intractable environmental problem. Eventually, this interest developed into an area of professional practice concerned with the process of mediation applied to such circumstances, and some of these environmentalists became full-time mediators.

This perspective differs from the first in that it is not particularly concerned with furthering a specific agenda or outcome in any decision-making process but emphasizes the ongoing design and improvement of such processes to make them more effective in dealing with multiple objectives drawn from different disciplines and participants with widely divergent views. The case study of British Columbia's Commission on Resources and Environment, by Owen, partially illustrates this perspective: the context is sustainable development, implying that environmental, social, and economic aspects of the issues are all substantively important, but the discussion focuses mainly on what worked in the process, rather than on its results.

Discussion

Sustainable development may have been the aim of this approach to policy integration as it developed, but new process-oriented fields, such as public consultation, facilitation, risk communication, and forms of conflict resolution like mediation and negotiations, have contributed heavily to its knowledge base. Research and insight are also drawn from traditional disciplines in the social sciences, such as social psychology, learning theory, and political science.

Probably the reaction from some readers to this process-based approach is to see it as shallow, without interest in content. This attitude, I think, misses both the approach's usefulness and its real intellectual significance.

In terms of usefulness, there is little question that the last two decades have seen a great increase not only in the number but also in the sophistication of processes designed to result in policies or decisions that more adequately integrate a multiplicity of perspectives. Some notable contributions to such processes have included

+ The work done on negotiation in the 1970s and 1980s at Harvard University and made widely available in the popular book *Getting to Yes* (Fisher and Ury 1986);

+ A multistakeholder consultation model based on the Niagara Institute's experience with labour–management relations and first introduced by Environment Canada in the mid-1980s;

+ Consensus decision-making principles developed as a joint project of Canadian national and provincial round tables (*Building Consensus for a Sustainable Future* [NRTEE 1993]); and

+ The founding and work of the first professional organization in the field of public consultation, now called the International Association for Public Participation, which by 1996 was engaged in a discussion of professional standards and ethics for this field.

Some of the obvious benefits of having experience, a growing body of research, and knowledgeable professionals in process issues include more effective participation and greater public confidence in the outcomes of such initiatives.

The intellectual contribution of this approach goes beyond these practical results. Just as with jury trials and democratic political systems, with integrated policy an emphasis on fair and inclusive processes, rather than on specific substantive results, is effectively a statement of faith in the citizen's responsibility for matters of public importance; it is also a statement about the nature of the results of trying to integrate several broad policy goals — sometimes surprising, sometimes disappointing, sometimes serendipitous, but always evolving. At no point is a permanent and final synthesis likely to be found; the problem of integration is inherently a dynamic one, and defining the idea of integrated policy, like the idea of democratic government, in terms of the qualities of its processes helps us to recognize that it is in a process of constant reformulation.

An assessment of strengths and weaknesses

As noted above, the contributions of this approach have been practical and have also provided some insights. However, process has not been widely used as the defining feature of policy integration; it has been much more common to treat process as a means to better substantive integration of economic, social, and environmental goals. Consequently, not much evaluation or even discussion has been made of integration as a process goal. Many questions arise about whether this approach is basically a matter of knowledge directed to practice and technique, like dentistry, or is really an established intellectual discipline, like political science. Perhaps most important is the question of whether it is worthwhile pursuing greater clarity about any of these issues. Is this approach an artificial category, useful for understanding themes in policy integration but not used enough to work with any further?

Policy integration as a substantive goal

The third perspective on policy integration considers the basic issues as being conceptual in nature; essentially, the problem is to adequately define sustainable development. This may at first glance seem to be a highly theoretical framework, with few practical implications, but I think that many people who take this approach would disagree. Indeed, returning to our historical perspective, the first task undertaken by most agencies newly created or newly mandated to promote or implement sustainable development has usually been to try to define the concept for themselves. Progress on such a broad goal as sustainable development cannot be assessed without an operational definition, and such practical matters as work programs require a clear articulation of what, exactly, they are trying to accomplish. Moreover, from the beginning of the public discussion of sustainable development, many people — social and environmental activists and academics in particular, but others, too — questioned whether the phrase *sustainable development* was actually meaningful or whether it masked such significant underlying contradictions that it was actually an oxymoron.

Discussion

Perhaps the most basic issue is whether this division of policy arenas or goals into three areas — social, economic, and environmental — is the most fruitful for discussing policy integration for sustainable development. Particularly in the context of measuring and reporting progress on sustainable development, researchers such as Hodge et al. (1995) prefer to formulate the problem in terms of the interaction of human activities with the environment and to identify policy goals as they relate to people, the environment, and the interactions between the human subsystem and its surrounding and supporting ecological systems. In such a model, the categorization of human activities as either economic or social is neither necessary nor very helpful. Many economically and socially necessary nonmarket activities, such as child care and home maintenance, are not easily categorized, so they may end up being poorly analyzed or simply ignored.

Another difficulty with using these three categories is that the boundaries between economic and social factors are unclear; economic factors can also be treated as a subset of the social. The economy and economic goals are not ends in themselves but are mechanisms for supplying human wants and needs. But people's wants and needs are mediated by the society they live in. And in every society, human beings create institutions related to law and governance, to personal bonds of affection and responsibility, and to meeting both material and nonmaterial needs, with all of these arrangements deriving from the biological fact that humans are a social species. Is the economy therefore not just a function of society?

There is, of course, much pragmatic value in using categories linked to identifiable academic disciplines and areas of public policy — at least the terms of the discussion are familiar (if somewhat fuzzy) and are clearly important. But the very fact that both academics and policymakers use these categories makes defining *integration* important. Does it mean trying to blend these subject areas in terms of the academic approaches used to describe and understand them? Some examples of this are attempts to redefine both national income and national wealth accounts in terms of resource, environmental, and heritage considerations; or attempts to recast economic theory on the basis of, or at least with reference to, choice and game theory or the laws of thermodynamics. But it seems quite unlikely that a unified theory of sustainable development will ever be developed or even attempted, because there are usually good reasons for dividing theory into limited but useful topic areas. What *policy integration* is most likely to mean, then, is either the development of compatible social, economic, and environmental goals or the development of compatible approaches to simultaneously achieve major goals in all three areas. These are not quite the same thing.

"The development of compatible ... goals" reflects the assumption that it is easier to modify what people want — what their goals are — than it is to achieve typical traditional goals in all three areas simultaneously. This is because social, economic, and environmental goals may in practice interfere with each other. In Robinson and Tinker's discussion of resocialization — by which they mean uncoupling human welfare from economic growth — they take essentially this approach. They discard the widely accepted and pursued goal of economic growth (personal income growth, at the individual level) as inappropriate for sustainable development, at least in the developed world and in the longer term, because it is too difficult to integrate with environmental quality goals. They substitute greater satisfaction from personal relationships and other social goods as more appropriate.

"The development of compatible approaches" reflects the assumption that using such means as new incentives and penalties, new technologies, and new institutional arrangements to address several goals simultaneously is easier than attempting to change people's important goals. This approach is exemplified by the small but growing number of companies that have improved environmental performance and worker safety while also improving their bottom line. Robinson and Tinker also support this approach in their comments on the usefulness of dematerialization and ecoefficiency as guides for sustainable development policy.

An assessment of strengths and weaknesses

I have already argued that clear thinking about policy integration and sustainable development has practical as well as theoretical value. The main weakness of this thinking, right now, is that there is not enough of it. The problem with much of the debate on sustainable development is the degree to which it is usually mired in ideology, rather than based on factual and theoretical investigation — or even on ordinary observation and common sense.

Simply not enough factual and analytical work is done to rigorously investigate all the assumptions in the discussion. Instead, the tendency has been to support a preexisting, ideologically based position. The hard questions in sustainable development arise only when seemingly indisputable but contradictory realities are juxtaposed. What kind of policy goals and approaches to pursuing them would emerge in a discussion that really tried to take full account of, for example,

+ The current understanding of the real ecological and environmental constraints to economic growth;

+ The implications of an apparently near-universal desire for cars and greater wealth;

+ The size of the human population and the preservation of biodiversity; and

+ The political, economic, and social barriers to making real progress on eliminating poverty?

Trying to investigate such issues and put the implications together in a coherent analysis is unquestionably a tall order. Perhaps such genuinely integrative policy questions are beyond the reasonable capabilities of any analyst. Nevertheless, it seems that even somewhat simpler issues — such as the role of economic growth in sustainable development, particularly in reducing poverty and inequity and improving the environmental impacts of technology — are not very much further developed than when the Brundtland Commission tackled them a decade ago.

Ignoring intractable realities, rather than wrestling with them, is too common a response. Certainly, one exception is the uncompromising insistence of some writers, such as Schrecker (1998), that both environmental problems and the impacts on poor people of some of the solutions be taken seriously. But the basic problem remains: too much is assumed about the substantive difficulties and incompatibilities involved in integrating policy goals and approaches in these areas, and too little is examined.

Conclusion

The implication that more work on the fundamental problems of policy integration and sustainable development may well reveal the futility of the whole effort is not really how I want to conclude this chapter. It may well be that human societies will face worsening social, environmental, and economic problems in the future — even catastrophes — as they did in the past. Nonetheless, the efforts to draw together goals and policies in these critical areas, to experiment with practical ways of bringing a more holistic view of the world to bear on policies and decisions, and to make decision-making processes more inclusive and more mindful of many aspects of reality are worthy initiatives — regardless of the state of the world. The tasks that the authors of this collection address can be seen as enormous (defining the policy directions of sustainable development) or quite limited (getting a single company to routinely pay attention to the environment). But the cumulative effects of many such efforts are the way, probably the only way, to effect deeply rooted change in society.

Contributing Authors

✧ ✦ ✧

Theo Anderson is currently Executive Director of Friends of the Earth, Ghana in Accra, Ghana. He has studied at the University of Science and Technology in Kumasi, Ghana, and the University College of London (UK), where he is now a PhD student in environmental economics and management. Mr Anderson's experience and expertise are in the areas of environmental assessment and evaluation; development policy and policy analysis; environmental and economic valuation of natural resources; and participatory rural appraisal.

Greg Armstrong is an international development consultant currently based in Southeast Asia. His areas of expertise include Southeast Asian governance, learning processes, education, human resource development, and institutional development. Dr Armstrong received his doctorate of education (EdD) from the University of Toronto (Canada) in 1981; he also holds an MA in educational planning (University of Toronto) and an MA in public administration (Carleton University, Ottawa, Canada).

Anne K. Bernard is a consultant specializing in education and international development, based in Ottawa, Canada. She holds a doctorate degree in education (EdD) and, for the past 25 years, has worked primarily in the development and monitoring of education programs and education research, and in conducting analyses and evaluations of social development programs.

Denis De Pape is a principal and senior consultant with InterGroup Consultants Ltd in Winnipeg, Canada. Mr De Pape is formally trained in natural resource and environmental economics and, for the past 25 year, has consulted for corporate, government, and nongovernment clients across western Canada. Mr De Pape is an instructor at the Natural Resources Institute of the University of Manitoba in Winnipeg. He also served as environmental policy analyst for Manitoba Hydro, where he contributed to the development of new environmental and sustainable development initiatives.

Susan Holtz is a private consultant and Adjunct Professor in the Environmental Planning Department of the Nova Scotia College for Art and Design. As a consultant, Ms Holtz specializes in energy, environment, and sustainable development policy, and works on related issues as a mediator and facilitator. She was founding Vice Chair of Canada's National, and Nova Scotia's, Round Table on the Environment and the Economy; a member of the Auditor General of Canada's Panel of Senior Advisors; and has served on the Canadian Environmental Advisory Council and the Canadian Environmental Assessment Research Council.

Ponciano S. Intal, Jr, since 1991, has been President of the Philippine Institute for Development Studies in Makati City, Philippines. He was previously Deputy Director-General of the National Economic and Development Authority for the Government of the Philippines. In 1983, Dr Intal received his PhD in economics from Yale University (New Haven, USA). His areas of expertise include international trade, development policy, and macroeconomic management.

Stephen Owen is Lam Professor of Law and Public Policy and Director of the Institute for Dispute Resolution at the University of Victoria (Canada). He is also a Commissioner of the Law Commission of Canada and the Canadian representative for the trilateral review of the North American Agreement on Environmental Cooperation. Professor Owen has previously been the Deputy Attorney General, Commissioner of Resources and Environment, Ombudsman, and Executive Director of the Legal Services Society of British Columbia. He has acted as an advisor to numerous international agencies on environmental, human rights, and conflict-resolution issues in Africa, Southeast Asia, Latin America, and Eastern Europe.

George Penfold manages a consulting practice in Comox, Canada, as an associate of the Westland Resource Group. Consulting activities include community and organizational planning and development; provincial, regional, and municipal resource use; and growth-management strategies. From 1991 to 1993, he was a member of the Commission on Planning and Development Reform in Ontario, and, from 1981 to 1995, he was a faculty member of the University School of Rural Planning and Development at the University of Guelph (Canada).

John Robinson is Director of the Sustainable Development Research Institute and Professor in the Department of Geography at the University of British Columbia in Vancouver, Canada. His current research interests include the reduction of large-scale greenhouse gas emissions; modeling sustainable futures in the Lower Fraser Basin of British Columbia; the interactions among sustainability, competitiveness, and trade; forecasting and futures studies; environmental philosophy; and the relationship between science and decision-making.

Jamie Schnurr is a Research Specialist at the International Development Research Centre in Ottawa, Canada. He holds degrees in environmental and development economics as well as in environmental policy and planning. Mr Schnurr's work has been primarily in the fields of waste management, reforestation, and community-based economic development. On behalf of IDRC, Mr Schnurr is currently developing a research program in the area of youth employment and sustainable livelihoods.

Luis Soberon A. is Senior Professor in the Department of Social Sciences of the Pontificia Universidad Catolica del Peru in Lima, Peru, as well as acting as a consultant for Escuela para el Desarrollo. Dr Soberon received his PhD in the sociology of development from Cornell University (Ithaca, USA). His principal interests include policies and programs for development, focusing on the study and analysis of institutions and social organizations, as well as project evaluation and strategic planning. Dr Soberon's current research focuses on the analysis of social structure and development as a cooperative effort.

Paul M. Syagga is Professor of Land Economics and Dean of the Faculty of Architecture, Design, and Development at the University of Nairobi (Kenya). Professor Syagga's research in land economics is widely published in both scholarly books and refereed academic journals. His research interests include real estate economics, housing policy, and environmental impact assessment. Professor Syagga is currently on sabbatical as a guest professor at the University of Pretoria (South Africa).

Jon Tinker is Senior Associate at the Sustainable Development Research Institute, University of British Columbia in Vancouver, Canada. Mr Tinker is also a Vancouver-based consultant and writer on international sustainable-development issues.

Vijay S. Vyas is Director of the Institute of Development Studies in Jaipur, India. Dr Vyas received hs PhD in economics from Bombay University in 1958. Dr Vyas' long career includes stints at the World Bank, as senior advisor in agriculture and rural development; IDRC, as senior fellow; Boston University, as visiting fellow; and the Indian Institute of Management in Ahmedabad (India), which he directed from 1978 until 1982.. He has chaired study teams convened by the Asian Development Bank, the World Bank, the Food and Agriculture Organization of the United Nations, and the International Fund for Agricultural Development, as well as serving on the board of the International Food Policy Research Institute. Dr Vyas has written extensively on various aspects of rural development and agricultural policy and has been honoured by the academic community in India and abroad for his contributions to these fields of research.

Abbreviations and Acronyms

✧ ✦ ✧

ADB	Asian Development Bank
AMO	Association of Municipalities of Ontario
APRA	Acción Popular Revolucionaria American (American popular revolutionary alliance) [Peru]
ASEAN	Association of Southeast Asian Nations
BATNA	best alternative to a negotiated agreement
BFAR	Bureau of Fishery and Aquatic Resources [Philippines]
CAC	command and control
CORE	Commission on Resources and Environment [British Columbia]
CPDR	Commission on Planning and Development Reform [Ontario]
CRB	community resource board
CTF	Community Tree Farming program [Philippines]
CVRP	Central Visayas Regional Project [Philippines]
DENR	Department of Environment and Natural Resources [Philippines]
EIA	environmental-impact assessment
ENRAP	Environment and Natural Resources Accounting Project [Philippines]
EPC	Environmental Protection Council [Ghana]
FIR	Frente Independiente Renovado (independent renovation front) [Peru]
FOE–Ghana	Friends of the Earth, Ghana
FOM	Forest Occupancy Management [Philippines]
GDP	gross domestic product
GERMP	Ghana Environmental Resources Management Project
IA	integrated assessment

IDRC	International Development Research Centre
IGADD	Inter-Governmental Authority on Drought and Development [Africa]
IGNB	*Indira Gandhi Nahar* Board [India]
IGNP	*Indira Gandhi Nahar Pariyojana* (Indira Gandhi Canal Project) [India]
IIMA	Indian Institute of Management, Ahmedabad
IRD	integrated rural development
ISFP	Integrated Social Forestry Program [Philippines]
MAF	million acre–feet
MBI	market-based instrument
MPP	Member of Provincial Parliament [Canada]
NCC	Nairobi City Council [Kenya]
NEAP	National Environmental Action Plan [Ghana]
NGO	nongovernmental organization
NIE	newly industrialized economy
OMB	Ontario Municipal Board
PARC	*Planning Act* Review Committee [Ontario]
PMA	pollution-management appraisal
SENATI	Servicio Nacional de Adestramiento en Trabajo Industrial (national technical training service) [Peru]
SIA	social-impact assessment
UPLB	University of the Philippines at Los Baños
UHP	Upland Hydroecology Program [University of the Philippines at Los Baños]
USAID	United States Agency for International Development
WFP	World Food Programme
WRI	World Resources Institute

Bibliography

❖ ◆ ❖

Acquasystems Consultants. 1989. Third Nairobi Water Supply Project: environmental action and monitoring programmes. Report to Nairobi City Council, Water and Sewerage Department, Nairobi, Kenya. 148 pp.

ADB (Asian Development Bank). 1990. Mangrove development project feasibility study. Vol. I. Final report. ADB, Mandaluyong City, Metro Manila, Philippines.

Agyepong, E.T. 1993. Wetlands management in Ghana. Friends of the Earth, Ghana, Accra, Ghana.

Akyuz, Y.; Chang, H.J.; Kozul-Wright, R. 1997. New perspectives on East Asian development. Paper distributed during the Workshop on Comparative African and East Asian Development Experience, 3–6 Nov 1997, Johannesburg, South Africa. African Economic Research Consortium, Nairobi, Kenya; United Nations University, Tokyo, Japan; Government of Japan, Tokyo, Japan; Korea Development Institute, Seoul, Korea. 42 pp.

Ali, R.M.; Rahim, K.A.; Nesadurai, H.; Hamid, N.A.; Mustapha, N.; Cheong, O.H.; Othman, R.; Aad, A.C. 1995. Trade and environment linkages: a Malaysian study. Report prepared for the United Nations Conference on Trade and Development, Geneva, Switzerland. Institute of Strategic and International Studies, Kuala Lumpur, Malaysia. 65 pp.

Allenby, B.R. 1992. Industrial ecology: the materials scientist in an environmentally constrained world. Materials Research Society Bulletin (Mar).

Alvarez, B. 1994. Factors affecting institutional success. In Alvarez, B.; Gomez, H., ed., Laying the foundation: institutions of knowledge in the developing countries. International Development Research Centre, Ottawa, ON, Canada. pp. 149–170.

Amat y León, C. 1994. Diagnóstico de la economía de la cuenca de Cajamarca. In Prioridades de investigacion y desarrollo en la Cuenca del Cajamarquino, documento base, Consorcio para el Desarrollo Sostenible de la Ecorregión Andina–Municipio de Cajamarca, Cajamarca, Peru. International Potato Center, Lima, Peru. pp. 21–93.

Amsden, A. 1994. Why isn't the whole world experimenting with the East Asian model to develop? Review of The East Asian miracle. World Development, 22(4), 627–633.

Anderson, R. 1993. AIDS: trends, predictions and controversy. Nature (London), 363, pp. 393–394.

Argyris, C.; Schon, D.A. 1978. Organizational learning: a theory of action perspective. Addison-Wesley, Don Mills, ON, Canada. 344 pp.

Bailey, P.D.; Gough, C.; Chapman, G.W. 1995. Methods for integrated environmental assessment: research directions for the European Union. Discussion paper presented at the European Commission Directorate General for Science, Research and Development XII/D-5 Workshop on Integrated Assessment, Stockholm, Sweden. Stockholm Environment Institute, Stockholm, Sweden.

Baldwin, S.; Cervinskas, J., ed. 1991. Community participation in research. Proceedings of a workshop, 23–27 Sep 1991, Nairobi, Kenya. International Development Research Centre, Ottawa, Canada. 57 pp.

BCSD (Business Council for Sustainable Development). 1994. Internalizing environmental costs to promote eco-efficiency. BCSD, Geneva, Switzerland.

Bennagen, P. 1978. Kaingin farming: a technical or sociological issue? *In* The search for a relevant framework for resource development: proceedings of a seminar–workshop. Program on Environmental Science and Management, University of the Philippines at Los Baños, Los Baños, Philippines.

Berman, P.; McLaughlin, M.W. 1976. Implementation of educational innovations. Educational Forum, 40(3), 344–370.

Bernard, A.K. 1991. Learning and intervention: the informal transmission of the knowledge and skills of development. *In* Perspectives on education for all. International Development Research Centre, Ottawa, ON, Canada. pp. 30–76.

Berrah, N. 1983. Energy and development: the tunnel effect. Revue de l'énergie, 356 (Aug–Sep).

Brown, L. 1981. Building a sustainable society. W.W. Norton, New York, NY, USA.

Bryson, J.M.; Crosbie, B.C. 1992. Leadership for the common good: tackling public problems in a shared power world. Jossey-Bass, San Francisco, CA, USA.

Cabrido, C.; Samar, E. 1994. Economic framework for Philippine land use decision-making: cost–benefit analysis of various land uses. *In* The Philippine Environment and Natural Resources Accounting Project (ENRAP phase II). Environmental Management Bureau, Department of Environment and Natural Resources, Quezon City, Philippines. pp. I1–I9.

Cadelina, R. 1986. Social forestry in the Philippine uplands: a university's perspective. *In* Rao, Y.; Hoskin, M.; Vergara, N.; Castro, C., ed., Community forestry: lessons from case studies in Asia and the Pacific region. East–West Center, Honolulu, HI, USA. pp. 103–134.

Castillo, R. 1983. How participatory is participatory development? A review of the Philippine experience. Philippine Institute for Development Studies, Metro Manila, Philippines.

CECARF (Consumer Energy Council of America Research Foundation). 1993. Incorporating environmental externalities into utility planning: seeking a cost-effective means of ensuring environmental quality. CECARF, Washington, DC, USA.

CGG (Commission on Global Governance). 1995. Our global neighbourhood. Report of the Commission on Global Governance. Oxford University Press, Oxford, UK.

Checkland, P.; Scholes, J. 1990. Soft systems methodology in action. John Wiley & Sons, New York, NY, USA.

CIVICUS (World Alliance for Citizen Participation). 1994. Citizens — strengthening global society. CIVICUS, Washington, DC, USA.

Clark, W.G.; Munn, R.E., ed. 1986. Sustainable development of the biosphere. International Institute for Applied Systems Analysis; Cambridge University Press, Cambridge, UK.

CMIE (Centre for Monitoring Indian Economy). 1992. Basic statistics relating to the Indian economy. Vol. 2: States. CMIE, Bombay, India.

———— 1994. Basic statistics relating to the Indian economy: states. CMIE, Bombay, India.

Cohen, S.J. 1996. Integrated assessment of climate change impacts: bringing it closer to home. Sustainable Development Research Institute, Vancouver, BC, Canada. SDRI Discussion Paper 96-2.

CORE (Commission on Resources and Environment). 1995. A provincial land-use strategy for British Columbia. CORE, Government of British Columbia, Victoria, BC, Canada.

Cote, R; Plunkett, T. 1996. Industrial ecology: efficient and excellent production. *In* Dale, A.; Robinson, J.B., ed., Achieving sustainable development. The University of British Columbia Press, Vancouver, BC, Canada. pp. 119–145.

CPDR (Commission on Planning and Development Reform in Ontario). 1993. New planning for Ontario: final report. CPDR, Toronto, ON, Canada.

Crombie, D. 1992. Regeneration: Toronto's waterfront and sustainability. Final report. Royal Commission on the Future of the Toronto Waterfront, Toronto, ON, Canada.

Cruz, M.; Zosa-Feranil, I.; Goce, C. 1988. Population pressure and migration: implications for upland development in the Philippines. Journal of Philippine Development, 15(1), 15–46.

Cruz, W.; Repetto, R. 1992. The environmental effects of stabilization and structural adjustment programs: the Philippine case. World Resources Institute, Washington, DC, USA.

Dale, A.; Robinson, J.; Massey, C. 1995. Reconciling human well-being and ecological carrying capacity: report on a series of workshops. Sustainable Development Research Institute, Vancouver, BC, Canada.

Daly, H. 1991. Steady-state economics (2nd ed., with new essays). Island Press, Washington, DC, USA.

Daly, H.; Cobb, J. 1994. For the common good — redirecting the economy toward community, the environment and a sustainable future (2nd ed.). Beacon Press, Boston, MA, USA.

Delors, J. 1996. Learning: the treasure within. United Nations Educational, Scientific and Cultural Organization, Paris, France. 189 pp.

de los Angeles, M.; Bennagen, M. 1993. Sustaining resource use in the Philippine uplands: issues in community-based environmental management. *In* Balisacan, A.; Nozawa, K., ed., Structures and reforms for rural development in the Philippines. Institute for Developing Economies, Tokyo, Japan.

de los Angeles, M.; Peskin, H. 1994. The Philippine Environment and Natural Resource Accounting Project. Phase two: framework and implementation strategy. Paper presented at the International Experts Meeting on Operationalization of the Economics of Sustainable Development, 28–30 Jul 1994, Manila, Philippines. Philippine Council for Sustainable Development, Pasig City, Metro Manila, Philippines; United Nations Commission on Sustainable Development, New York, NY, USA.

Dobell, R. 1994. The "dance of the deficit" and the real world of wealth: re-thinking economic management for social purpose. School of Public Administration, University of Victoria, Victoria, BC, Canada. Draft.

Doering, R.; Biback, D.M.; Muldoon, P.; Richardson, N.; Rust-D'Eye, G.H. 1991. Planning for sustainability: towards integrating environmental protection into land use planning. Royal Commission on the Future of the Toronto Waterfront, Toronto, ON, Canada.

Drucker, P.F. 1995. Really reinventing government. Atlantic Monthly, 275(2).

Dunlap, R.E.; Gallup, G.H., Jr; Gallup, A.M. 1993. Health of the planet: results of a 1992 international environmental opinion survey of citizens in 24 nations. George H. Gallup International Institute, Princeton, NJ, USA.

Durning, A. 1992. How much is enough? The consumer society and the future of the Earth. W.W. Norton, New York, NY, USA.

Ekins, P.; Jacobs, M. 1994. Are environmental sustainability and economic growth compatible? Department of Applied Economics, University of Cambridge, Cambridge, UK. Energy–Environment–Economy Modelling Discussion Paper No. 7.

Elkins, D. 1995. Beyond sovereignty — territory and political economy in the twenty-first century. University of Toronto Press, Toronto, ON, Canada.

EMB (Environmental Management Bureau). 1990. The Philippine environment in the eighties. Department of Environment and Natural Resources, Quezon City, Metro Manila, Philippines.

Etzioni, A. 1993. The spirit of community — the reinvention of American society. Simon and Schuster, Toronto, ON, Canada.

FAO (Food and Agriculture Organization of the United Nations). 1982. Management and utilization of mangroves in Asia and the Pacific. FAO, Rome, Italy.

Finger, M; Verlaan, P. 1995. Learning our way out: a conceptual framework for socio-environmental learning. World Development, 23(3), 503–513.

Fisher, R.; Ury, W. 1986. Getting to yes. Business Books Ltd, London, UK. 161 pp.

FOE–Ghana (Friends of the Earth, Ghana). 1994. Wetlands management in Ghana — inventory classification and distribution. FOE–Ghana, Accra, Ghana. 58 pp.

Frias, C. 1993. Diagnostico socio-productivo del Departamento de Cajamarca. Intermediate Technology Development Group, Lima, Peru. Working document. 98 pp.

Friedman, J. 1987. Planning in the public domain: from knowledge to action. Princeton University Press, Princeton, NJ, USA.

Friend, J.; Hickling, J. 1987. Planning under pressure: the strategic choice approach. Pergamon Press, London, UK.

Frosch, R.A.; Gallapoulous, N.E. 1992. Towards an industrial ecology. In Bradshaw, A.D.; Southwood, R.; Warner, F., ed., The treatment and handling of wastes. Chapman and Hall, for the Royal Society, London, UK.

Fullan, M. 1993. Change forces: probing the depths of educational reform. The Falmer Press, London, UK. 162 pp.

———— 1996. Leadership for change. In Leithwood, K.; et al., ed., International handbook of educational leadership and administration. Kluwer Academic, Dordrecht, Netherlands. pp. 701–722.

Funtowicz, S.O.; Ravetz, J.R. 1993. Science for the post normal age. Futures, 25(7), 739–755.

GBC (Government of British Columbia). 1992. Commissioner on Resources and Environment Act. Government of British Columbia, Victoria, BC, Canada.

Glover, D. 1994. Policy researchers and policy makers: never the twain shall meet? Journal of Philippine Development, 21(1–2), 63–89.

GMP (Government of Madhya Pradesh). 1995. Madhya Pradesh human development report. GMP, Bhopal, India.

GOK (Government of Kenya). 1968. The laws of Kenya. Chapter 295: The Land Acquisition Act. Government Printers, Nairobi, Kenya.

Goldemberg, J. 1992. Technological leapfrogging. Loyola of Los Angeles International and Comparative Law Journal, 15(1).

GOR (Government of Rajasthan). 1992. Trends in land use statistics. Directorate of Agriculture (Statistical Wing), GOR, Jaipur, India.

Guha, R. 1989. Radical American environmentalism and wilderness preservation: a Third World critique. Environmental Ethics, 11 (Spring), 71–83.

Gustavson, K.R.; Lonergan, S.C. 1994. Sustainability in British Columbia: the calculation of an index of sustainable economic well-being. Centre for Sustainable Regional Development, Victoria, BC, Canada.

Hashim, S.R. 1995. Development and employment. Indian Journal of Labour Economics, 38(1), 41–53.

Hawken, P. 1993. The ecology of commerce — a declaration of sustainability. Harper Business, New York, NY, USA.

Hodge, T.; Holtz, S.; Smith, C.; Baxter, K.H., ed. 1995. Pathways to sustainability: assessing our progress. National Round Table on the Environment and the Economy, Ottawa, ON, Canada. 229 pp.

Holling, C. 1986. The resilience of terrestrial ecosystems: local surprise and global change. In Clark, W.; Munn. R., ed., Sustainable development of the biosphere. Cambridge University Press, Cambridge, UK.

Homer-Dixon, T. 1991. On the threshold: environmental changes as causes of acute conflict. International Security, 16(2), 76–116.

Hooja, R. 1994. Colonization and settler motivation efforts. In Hooja, R.; Kavadia, P.S., ed., Planning for sustainability in irrigation. Rawat Publication, Jaipur, India. pp. 163–192.

Howard Humphreys (Kenya) Ltd; Environmental Resources. 1988. Third Nairobi Water Supply Project: environmental assessment study. Report to Nairobi City Council, Water and Sewerage Department, Nairobi, Kenya. 96 pp.

Hunter, J.M. 1967. Population pressure in part of the West African savanna. A study of Mangodi in northeast Ghana. Annals of the Association of American Geographers, 67, 101–114.

IDA (Irrigation Development Authority). 1990. Progress report 6: Technical assistance twinning of GIDA and ADC. Ministry of Agriculture, Accra, Ghana.

Ignatieff, M. 1993. Blood and belonging — journeys into the new nationalism. Penguin, Toronto, ON, Canada.

IGNB (Indira Gandhi Nahar Board). 1995. Profile of Indira Gandhi Nahar Project. Government of Rajasthan, Jaipur, India. pp. 10–15.

IGNP (Indira Gandhi Nahar Pariyojana). 1996. Project profile. Government of Rajasthan, Jaipur, India.

IISD (International Institute for Sustainable Development). 1994. Green budget reform: casebook on leading practices. IISD, Winnipeg, MB, Canada.

Intal, P. 1995. Comments on *The East Asian miracle: economic growth and public policy*. In The World Bank's East Asian miracle report: its strengths and limitations. Overseas Economic Cooperation Fund; Research Institute of Development Assistance, Tokyo, Japan. OECF Discussion Paper 7. pp. 164–166.

Intal, P.; Israel, D.; Quintos, P.; de los Angeles, M.; Medalla, E.; Pineda, V.; Tan, E.; Pelayo, R. 1994. Trade and environment linkages: the case of the Philippines. Report prepared for the United Nations Conference on Trade and Development, Geneva, Switzerland. Policy and Development Foundation, Inc., Manila, Philippines.

IPCC (Intergovernmental Panel on Climate Change). 1998. Report of Working Group III — the socio-economic dimensions of climate change. Cambridge University Press, Cambridge, UK.

IRG (International Resources Group)–Edgevale Associates. 1991. The Philippine Environment and Natural Resources Accounting Project — phase one: executive summary. Report submitted to Department of Environment and Natural Resources, Quezon City, Philippines; United States Agency for International Development, Washington, DC, USA.

———— 1994. Philippine Environmental and Natural Resources Accounting Project — phase two: main report. Report submitted to Department of Environment and Natural Resources, Quezon City, Philippines; United States Agency for International Development, Washington, DC, USA.

IUCN (International Union for the Conservation of Nature). 1980. World conservation strategy. IUCN, Gland, Switzerland.

Jodha, N.S. 1990. Rural common property resources: contribution and crisis. Society for Promotion of Wasteland Development, New Delhi, India. Foundation-Day Lecture.

Joseph, J. 1997. The round table for consensus building in Cajamarca, Peru. Grassroots Development, 21(1), 40–45.

JVM (Juntos Vecinos y Municipalidad). 1994. Memoria 1993 [annual report 1993]. JVM, Cajamarca, Peru. 80 pp.

Kanoh, T. 1992. Toward dematerialization. In International Institute for Applied Systems Analysis, ed., Science and sustainability — selected papers on IIASA's 20th anniversary. International Institute for Applied Systems Analysis, Laxenburg, Austria. pp. 63–94.

Kapetsky, J.M. 1991. Framework for culture-based fisheries in Ghana. Food and Agriculture Organization of the United Nations, Rome, Italy. Field Technical Report 1.

Kay, J.; Schneider, E. 1994. Embracing complexity: the challenge of the ecosystem approach. Alternatives, 20(3), 32–39.

Kiriro, A.; Juma, C. 1989. Gaining ground: institutional innovations in land use management in Kenya. African Centre for Technology Studies, Nairobi, Kenya. 186 pp.

Korten, F.; Siy, R., ed. 1988. Transforming a bureaucracy. Kumarian Press, West Hartford, CT, USA.

Kuik, O.; Verbruggen, H. 1991. In search of indicators for sustainable development. Kluwer Academic, Dordrecht, Netherlands. 231 pp.

Laing, F., ed. 1994. Ghana environmental action plan, volume 2. Environmental Protection Council, Accra, Ghana. 325 pp.

Lee, K.N. 1993. Compass and gyroscope: integrating science and politics for the environment. Marvyn Melnyk Associates, Queenston, ON, Canada. 235 pp.

Lindquist, E.A. 1992. Public managers and policy communities: learning to meet new challenges. Canadian Public Administration, 35(2), 127–159.

Lynn, C.W. 1945. Land planning in the northern territories of Ghana. Farm and Forests, 7.

Mamphey, K.; Agyei, O.A. 1985. Statement on planning and management of human settlements with emphasis on small and intermediate towns and local growth points. Paper presented at the Conference on Planning and Management of Human Settlements, Mar 1988, Kingston, Jamaica. United Nations Centre for Human Settlements (Habitat), Nairobi, Kenya.

Mann, J.; Tarantola, D.; Netter, T.W., ed. 1992. AIDS in the world. Harvard University Press, Cambridge, MA, USA.

Mariano, S. 1986. Participatory upland development: views from Rizal, Philippines. In Rao, Y.S.; Hoskin, M.; Vergara, N.; Castro, C., ed., Community forestry: lessons from case studies in Asia and the Pacific region. East–West Center, Honolulu, HI, USA.

Marshall, J.; Senkiw, C; Penfold, G. 1995. Key informant perspectives on planning and development reform in Ontario. International Development Research Centre, Ottawa, ON, Canada.

Mason, A. 1997. Population change in East Asia's miracle economies: is there a connection? Paper presented at the Policy Seminar on African and Asian Economic Development: Long Term Perspective, 20–21 Oct 1997, Tokyo, Japan. Nihon University, Tokyo, Japan; East–West Center, Honolulu, HI, USA; World Bank, Washington, DC, USA.

Mathur, H.M. 1991. Planners and the poor in Indira Gandhi Canal region: anthropological perspective on the development of new rural settlements. In Mathur, P.C.; Gurjar, R.K., ed., Water and land management in arid ecology. Rawat Publication, Jaipur, India. pp. 241–260.

Mathur, P.C.; Gurjar, R.K. 1991. Water and land management in arid ecology. In Sharma, S.; Saiwal, S., ed., Socio-economic and ecological dimensions of arid land in Rajasthan. Rawat Publication, Jaipur, India.

MDC (Municipalidad de Cajamarca). 1994. Concertación institucional y plan de desarrollo [institutional concertation and development plan]. Asociación para la Sobrevivencia y el Desarrollo Local; Centro Ecunénico de Promoción y Acción Social; Equipo de Desarrollo Agropecuario de Cajamarca–Centro de Investigación Educación y Desarrollo. MDC, Cajamarca, Peru. 74 pp.

Mendoza, N. 1994. Input–output modeling. In The Philippine Environment and Natural Resources Accounting Project (ENRAP phase II). Environmental Management Bureau, Department of Environment and Natural Resources, Quezon City, Philippines. pp. G1–G52.

Mesa de Concertación. 1995. Proceedings, Cajamarca: Democracia, Medio Ambiente y Desarrollo. Seminar, 11–13 Jan 1995, City of Cajamarca, Cajamarca, Peru. Mesa de Concertación. Thematic Round Tables Preliminary Documents.

Metcalfe, L. 1993. Public management: from imitation to innovation. In Koom, J., ed., Modern governance: new government–society interactions. Sage, London, UK. pp. 173–286.

Michael, D.N. 1992. Governing by learning in an information society. In Rosell, S., Governing in an information society. Institute for Research on Public Policy, Montréal, PQ, Canada. pp. 121–133.

Mintzberg, H. 1989. Mintzberg on management. The Free Press, New York, NY, USA. 256 pp.

Morley, D. 1989. Frameworks for organizational change: towards action learning in global environments. In Wright, S.; Morley, D., ed., Learning works: searching for organizational futures. ABL Group, Faculty of Educational Studies, York University, Toronto, ON, Canada.

Moull, T. 1993. Consolidated research report. Commission on Planning and Development Reform in Ontario, Toronto, ON, Canada.

Mujica, E. 1995. The "concertation table" of Cajamarca (Perú): a case study on integrated policy at the local government level in Latin America. Paper presented at the IDRC Round Table Workshop on Integrating Social, Economic and Environmental Policy: A Matter of Learning, 8–9 May 1995, Ottawa, ON, Canada. International Development Research Centre, Ottawa, ON, Canada. 16 pp.

Najam, A. 1995. Learning from the literature on policy implementation: a synthesis perspective. International Institute for Applied Systems Analysis, Laxenburg, Austria. Working Paper. 70 pp.

NCKP (National Conference on the Kaingin Problem). 1964. Conference recommendations. College of Forestry, University of the Philippines at Los Baños, Los Baños, Philippines. 84 pp.

NDPC (National Development Planning Commission). 1991. Making people matter — a human development strategy for Ghana. Government of Ghana, Accra, Ghana.

———— 1995. National development policy framework — a twenty-five year perspective for the transformation of the Ghanaian economy. Government of Ghana, Accra, Ghana.

North, D. 1994. Economic performance through time. American Economic Review, 84(3), 359–368.

NRTEE (National Round Table on the Environment and the Economy). 1993. Building consensus for a sustainable future: guiding principles. NRTEE, Ottawa, ON, Canada. 22 pp.

O'Connor, D. 1994. Managing the environment with rapid industrialization: lessons from the East Asian experience. Organisation for Economic Co-operation and Development, Paris, France. Development Centre Studies Series.

OEC (Ontario Economic Council). 1971. Subject to approval: a review of municipal planning in Ontario. Ontario Economic Council, Toronto, ON, Canada.

Ontario. 1989. *Planning Act* 1983, statutes of Ontario. Ontario Ministry of the Attorney General, Toronto, ON, Canada.

———— 1994. Bill 193, *An Act to Amend the Planning Act*. Ontario Ministry of the Attorney General, Toronto, ON, Canada.

O'Riordan, T. 1991. Towards a vernacular science of environmental change. *In* Weale, A.; Roberts, L.; Saunders, P., ed., Innovation, technology and society. Belhaven Press, London, UK. pp. 143–156.

———— 1994. Civic science and global environmental change. Scottish Geographical Magazine, 110(1), 4–12.

Pace University Center for Environmental Legal Studies. 1990. Environmental costs of electricity. Oceana Publications, New York, NY, USA.

Page, J. 1994. The East Asian miracle: building a basis for growth. Finance and Development (Mar), 2–5.

PARC (*Planning Act* Review Committee). 1977. Report of the *Planning Act* Review Committee. PARC, Toronto, ON, Canada.

Paul, S. 1992. Accountability in public services: exit, voice and control. World Development, 20(7), 1047–1060.

Pauly, D.; Christensen, V. 1995. Primary production required to sustain global fisheries. Nature (London), 374, 255–257.

Pearce, D. 1994. Sustainable consumption through economic instruments. *In* Norwegian Ministry of Environment, ed., Sustainable consumption, report of a symposium held in Oslo, Norway, 19–20 Jan 1994. Norwegian Ministry of Environment, Oslo, Norway. pp. 84–90.

Penalba, L.; de los Angeles, M.; Francisco, H. 1994. Impact evaluation of the Central Visayas Regional Project phase 1 (CVRP-1). Philippine Institute for Development Studies, Makati City, Metro Manila, Philippines. Discussion Paper 94-22. 240 pp.

Phantumvanit, D.; Wigzell, S; Buranakul, E.; Bootherawara, N.; Bowonwiwat, R. 1994. The interlinkages between trade and environment: Thailand. Report prepared for the United Nations Conference on Trade and Development, Geneva, Switzerland; Thailand Environment Institute, Bangkok, Thailand. 175 pp.

Pressman, J.L.; Wildavsky, A. 1973. Implementation. University of California Press, Berkeley, CA, USA. 304 pp.

Ramanathan, S.; Rathore, M.S. 1991. Equity and productivity issues in IGNP — a review of literature. Institute of Development Studies, Jaipur, India. Mimeo.

———— 1994. Sustainability of Indira Gandhi Canal and the need for correct responses: a social scientist's perspective, *In* Hooja, R.: Kavadia, P.S., ed., Planning for sustainability in irrigation. Rawat Publication, Jaipur, India. pp. 277–299.

Ranis, G. 1995. Another look at the East Asian miracle. World Bank Economic Review, 9(3), 509–534.

RCEUDPC (Royal Commission on the Economic Union and Development Prospects for Canada). 1985. Report of the Royal Commission on the Economic Union and Development Prospects for Canada. Minister of Supply and Services, Ottawa, ON, Canada.

Rees, W. 1995. More jobs, less damage: a framework for sustainability, growth and employment. Alternatives, 21(4), 24–30.

Rees, W.; Wackernagel, M. 1994. Ecological footprints and appropriated carrying capacity: measuring the natural capital requirements of the human economy. *In* Jansson, A., ed., Investing in natural capital — the ecological economics approach to sustainability. Island Press, Washington, DC, USA.

Regier, H. 1995. The limits of ecological carrying capacity. *Summarized in* Dale, A.; Robinson, J.; Massey, C. Reconciling human well-being and ecological carrying capacity: report on a series of workshops. Sustainable Development Research Institute, University of British Columbia, Vancouver, BC, Canada.

Repetto, R.; Magrath, W.; Wells, M.; Beer, C.; Rossini, F. 1989. Wasting assets: natural resources in the national income accounts. World Resources Institute, Washington, DC, USA. 68 pp.

Rifkin, S. 1996. Paradigms lost: toward a new understanding of community participation in health programmes. Acta Tropica, 61, 79–92.

Robins, N. 1993. How business can contribute to sustainable development: getting eco-efficient. Report of the Business Council on Sustainable Development 1st Antwerp Eco-efficiency Workshop. International Institute for Environment and Development, London, UK.

Robinson, J. 1988. Unlearning and backcasting: rethinking some of the questions we ask about the future. Technological Forecasting and Social Change, 33(4), 325–338.

———— 1991. Modelling the interactions between human and natural systems. International Social Sciences Journal, 130, 629–647.

———— ed. 1998. Life in 2030 — exploring a sustainable society in Canada. The University of British Columbia Press, Vancouver, BC, Canada. (In press.)

Robinson, J.; Timmerman, P. 1993. Myths, rules, artifacts, ecosystems: framing the human dimensions of global change. *In* Wright, S.D.; Dietz, T.; Borden, R.; Young, G.; Guagnano, G., ed., Human ecology: crossing boundaries. Selected papers from the 6th Conference of the Society for Human Ecology, 2–4 Oct 1992, Snowbird, UT, USA. Society for Human Ecology, Bar Harbor, ME, USA. pp. 236–246.

Robinson, J.; Van Bers, C. 1996. Living within our means: the foundations of sustainability. David Suzuki Foundation, Vancouver, BC, Canada.

Rola, W.R. 1987. Impact assessment of the Jala-Jala agroforestry area: a multidimensional approach. *In* Ponce, E.; Salazar, R.; Zerudo, A., ed., Sustaining upland development. Manila Research Center, De La Salle University, Manila, Philippines.

Rosell, S. 1995. Changing maps: governing in a world of rapid change. Carleton University Press, Ottawa, ON, Canada. 253 pp.

Rothman, D.; Robinson, J. 1997. Growing pains: towards a conceptual framework for considering integrated assessments. Sustainable Development Research Institute, Vancouver, BC, Canada. SDRI Discussion Paper 96-1.

Roy, T.K. 1983. Impact of Rajasthan Canal Project on social, economic and environmental conditions. National Council of Applied Economic Research, New Delhi, India.

Sachs, I. 1984. Strategies for ecodevelopment. Ceres, 17(4).

Sagasti, F.; Patron, P.; Lynch, N.; Hernandez, M. 1995. Agenda Peru: democracy and good government. Perú Apoyo, Lima, Peru. 149 pp.

Sain, K. 1978. Reminiscences of an engineer. Young Asia Publications, New Delhi, India; Stockholm, Sweden.

Sajise, P. 1984. Strategies for transdisciplinary research on ecosystem management: the case of the UPLB Upland Hydroecology Program. In Rambo, T.; Sajise, P., ed., An introduction to human ecology research on agricultural systems in Southeast Asia. University of the Philippines at Los Baños, Los Baños, Philippines; East–West Center, Honolulu, HI, USA. pp. 312–327.

Sanders, H.D. 1992. The Khazzoom–Brookes postulate and neoclassical growth. Energy Journal, 13(4), 131–148.

Schmidt-Bleek, F. 1992a. MIPS — a universal ecological measure. Fresenius Environmental Bulletin, 1, 306–311.

———— 1992b. MIPS revisited. Fresenius Environmental Bulletin, 2, 407–412.

———— 1993. Toward universal ecology disturbance measures: basis and outline of a universal measure. Regulatory Toxicology and Pharmacology, 18(3).

Schrecker, T. 1998. Sustainability, growth and distributive justice: questioning environmental absolutism. In Lemons, J.; Westra, L.; Goodland, R., ed., Environmental sustainability and integrity. Kluwer, Dordrecht, Netherlands. (In press.)

Seetharaman, S.P.; Barua, S.K.; Jajoo, B.H.; Singhi, P.M. n.d. Indira Gandhi Nahar Project (IGNP) — an approach to increase rate of settlement in phase II. Indian Institute of Management, Ahmedabad, India. 54 pp.

Shackley, S.; Wynne, B. 1995. Integrating knowledges for climate change: pyramids, nets and uncertainties. Global Environmental Change, 5(2), 113–126.

Sharma, R.; Rathore, M.S. 1990. A report on the impact of IGNP on labour and employment. Institute of Development Studies, Jaipur, India. Mimeo.

Simpson, R.D. 1995. Environmental policy, innovation and competitive advantage. Resources for the Future, Washington, DC, USA. RFF Discussion Paper 95-12.

Singh, S. 1994. Credit system in a physically hostile environment — a case of new settlers in IGNP area. Institute of Development Studies, Jaipur, India. Mimeo.

Sloan, G. 1945. Report of the Commissioner Relating to the Forest Resources of British Columbia. Government of British Columbia, Victoria, BC, Canada.

Sopuck, R.D. 1993. Canada's agricultural and trade policies: implications for rural renewal and biodiversity. National Round Table on the Environment and the Economy, Ottawa, ON, Canada. Working Paper No. 19. 51 pp.

Srivastava, K.; Rathore, M.S. 1992. People's initiative for development: Indira Gandhi Nahar Yatra. Economic and Political Weekly, 27(9), 450–454.

Sterkenburg, J.J. 1987. Rural development and rural development policies: cases from Africa and Asia. Netherlands Geographical Studies, Amsterdam, Netherlands. 257 pp.

Syagga, P.M. 1994. Promoting sustainable construction industry activities in Kenya. African Centre for Technology Studies, Nairobi, Kenya. Research Memorandum No. 8. 17 pp.

Syagga, P.M.; Olima, W.H. 1996. Impact of land acquisition on displaced households: case study of Third Nairobi Water Supply Project. Habitat International, 20(1), 61–75.

Syagga Associates. 1988. Third Nairobi Water Supply Project: socio-economic survey. Report to Nairobi City Council, Water and Sewerage Department, Nairobi, Kenya. 57 pp.

Tan, J.A.; Intal, P. 1992. Sustainable development in development planning and policy-making: the Philippine case. Canberra Bulletin of Public Administration, 69, 114–124.

Thomas, A. 1989. Learning communities and the cultural implications of global learning. Ontario Institute for Studies in Education, Toronto, ON, Canada. Research Paper.

Tibbs, H.B.C. 1992. Industrial ecology — an agenda for environmental management. Pollution Prevention Review (Spring), 167–180.

Tolentino, B. 1991. Promoting forestry as a land use under the Philippines Social Forestry Programme. In Mathoo, M.; Chipeta, M., ed., Trees and forests in rural land use. Forestry Department, Food and Agriculture Organization of the United Nations, Rome, Italy.

Trevallin, B.A.W. 1994. Planning in the twenty-first century: prospects and challenges. Keynote address to the Conference on the Planner as an Agent of Change, 18 Aug 1994, Accra, Ghana. Ghana Institute of Planners, Accra, Ghana.

Trevallin, B.A.W.; Tetteh, O. 1976. Organisation for planning in Ghana. Town and Country Planning Department, Accra, Ghana.

Turner, B.; Clark, W.; Kates, R.; Richards, J.; Matthews, J.; Meyen, W. 1990. The Earth as transformed by human action. Cambridge University Press, Cambridge, UK.

Turner, B.L.; Benjamin, P.A. 1994. Fragile lands: identification and use for agriculture. In Ruttan, V.W., ed., Agriculture, environment and health: sustainable development in the 21st century. University of Minnesota Press, Minneapolis, MI, USA; London, UK.

UNCED (United Nations Conference on Environment and Development). 1992. Agenda 21: programme of action for sustainable development. United Nations, New York, NY, USA.

UNDP (United Nations Development Programme). 1994. Human development report, 1994. Oxford University Press, New York, NY, USA.

UNICEF (United Nations Children's Fund). 1994. The state of the world's children — 1994. Oxford University Press, Oxford, UK.

Villarán, S. 1994. Los desposcidos. La Republica, 14 Aug, Sunday Supplement.

Vitousek, P.; et al. 1986. Human appropriation of the products of photosynthesis. BioScience, 34(6), 368–73.

von Weizsacker, E. 1994. How to achieve progress towards sustainability. In Norwegian Ministry of Environment, ed., Sustainable consumption, report of a symposium held in Oslo, Norway, 19–20 Jan 1994. Norwegian Ministry of Environment, Oslo, Norway.

WAPCOS (Water and Power Development Consultancy Services). 1989. Ecological and environmental studies for integrated development of Indira Gandhi Nahar Project, stage II — draft report. WAPCOS, New Delhi, India.

———— 1992a. Ecological and environmental studies for integrated development of Indira Gandhi Nahar Project, stage II — final report. WAPCOS, New Delhi, India.

———— 1992b. Indira Gandhi Nahar Project, stage II — draft final feasibility report. Vol. 1. WAPCOS, New Delhi, India. 17 pp.

WCED (World Commission on Environment and Development). 1988. Our common future. Oxford University Press, Oxford, UK.

World Bank. 1992. World development report 1992. Oxford University Press, New York, NY, USA.

———— 1993. The East Asian miracle: economic growth and public policy. Oxford University Press, New York, NY, USA. 389 pp.

———— 1994. World development report 1994. Oxford University Press, New York, NY, USA.

WI (Worldwatch Institute). 1995. State of the world, 1995. W.W. Norton, New York, NY, USA.

Williams, R.; Larson, E.; Ross, M. 1987. Materials, affluence and industrial energy use. Annual Review of Energy, 12, 99–144.

WRI (World Resources Institute). 1994. World resources 1994–95. Oxford University Press, Oxford, UK.

Wright, S.; Morley, D., ed. 1989. Learning works: searching for organizational futures. ABL Group, Faculty of Educational Studies, York University, Toronto, ON, Canada. 292 pp.

Wynne, B. 1992. Uncertainty and environmental learning: reconceiving science and policy in the preventive paradigm. Global Environmental Change, 2 (June), 111–127.

Young, J.; Sachs, A. 1994. The next efficiency revolution: creating a sustainable materials economy. Worldwatch Institute, Washington, DC, USA. Worldwatch Paper 121.

Zandstra, H.; Swanberg, K.; Zulberti, C.; Nestel, B., ed. 1979. Caqueza: living rural development. International Development Research Centre, Ottawa, ON, Canada. 321 pp.

Date Due

MY 29 03			